Red Flag:
From the Ground Up

A first-ever walk around of Red Flag's renowned threat range with the Captain who made it tick.

By Kernan Chaisson

(Photo: U.S. Air Force)

DEFENSE
LION
PUBLICATIONS

Dedication

**To Military Families,
the heroes
who support the heroes.**

**To Nadine, my hero
Thanks for the support and patience.**

Previous book by author

*MAD Cats:
The story of VP-63*

Contents

Contents

Contents

Preface

Writing in *Air Force Magazine,* Walter J. Boyne did a special piece for the 25th Anniversary of the world famous Red Flag combat training exercise. He closed with the following:

> *One major milestone in that history, without question, was the stunning performance of American airmen in the Gulf War of 1991. It was the first war to showcase the results of Red Flag, and it produced a curious tribute. It came from an Air Force pilot who, returning from a combat mission over Iraq, was heard to remark, "It was almost as intense as Red Flag."*
> *Air Force Magazine,* November, 2000.

The United States Air Force is second to none. When combined with select allies airspace dominance is a given. Large numbers of the best aircraft are part of this equation; but knowing how to fully use their capabilities and understanding how to work together does not come without training. Since the creation of Red Flag, Nellis AFB and its range complex has become the epicenter of Air Force fighter pilot training.

Red Flag takes advantage of the 6,000 square miles of flyable, restricted air space left over from weapons testing during and after World War II. Pilots practice air-to-air combat with "Aggressor" aircraft which simulate enemy tactics. Unlike the smaller Navy Top Gun program in California, Red Flag became a large, multi-squadron exercise where fliers could hone individual combat skills while working together to accomplish a specific mission. It increasingly became international in scope.

Building on its success in air-to-air training, Red Flag began to expand by adding ground-to-air threats so pilots could develop their skills with the new self-protection electronic warfare systems being installed on their airplanes, skills that would increase survival possibilities in a multi-threat environment.

Recognizing the opportunity that was presenting itself, Air Force officials began growing the air combat training environment and establishing a simulated air defense system to challenge pilots as they tried to accomplish their missions. Nellis and the desert range complex half the size of Switzerland had long been a special place for developing and testing all manner of weapons and munitions. Why not build an enemy country to attack over and over?

Viet Nam was over, so the natural tendency was to pattern the "enemy" on the Soviet Union's Integrated Air Defense System (IADS). This included early warning, anti-aircraft networks, gun-laying radars, as well as missile tracking sensors. The goal was to insure that the first time a pilot saw this environment was not on his or her first bomb run on a Fulda Gap (in those days) or any real target in real combat.

From a way of giving pilots the air-to air combat practice they needed to survive, Red Flag became the environment where squadrons could practice for war as a team; a war they hoped would never come, but would be ready to for if it ever did.

The threat systems were engineered so the electronic warning systems of fighters "saw" SA-2, SA-3, SA-6, and ZSU-23-4 or early warning radars. Planners saw the value of providing a way to debrief pilots on what they did right and what they did wrong. They then were given a chance to try again, and again, until they got it right.

A combat training environment as complex and challenging as Red Flag is more than buildings, runways, radars, and radios. Planning is important; but people are the critical juice that makes it all work

Over the years, attention increased on the unique world of Nellis and Red Flag – especially the planes. Books, videos, there even was a movie that told about Red Flag and the exciting world of the fighter pilot. But while the jets roared about the sky getting all the attention, 'converting JP-5 to noise' as we used to say, crews of hard-working, skilled operators and technicians toiled in the desert dust below, miles and miles from the base and home, to run the threat simulator systems that provided pilots with life-saving training. This made Nellis and Red Flag one of the most unique and valuable combat training experiences on the planet.

This book covers the period from late 1979 to early 1985. The ground range was expanding and improving. It was perched on the edge of a major expansion of equipment and mission.

Given today's experience with a decade of endless war, it may be worthwhile to look back at another time, another life; maybe not better, but definitely different. We could focus on getting good at a skill, not worrying about actual survival. Pilots got to practice in an environment where death was not the companion of mistakes.

Frank Bruni, in the 15 September 2013 *New York Times* wrote that over 2 million troops deployed to Iraq and Afghanistan. There were 6,500 dead and 1,500 vets are amputees and 20 to 30% are carrying the darkness and personality changes of PTSD. The reader may find it interesting to glimpse a military life not filled with such misery and hurt. This is not to be confused with a lack of seriousness. But the force did not have a sense of futility because the leftovers of the Cold War (the Berlin Wall had not yet come down) made everyone feel that proof of skill and effectiveness is what insured a lack of war, a big change from today. For the Red Flag staff a joy of what we did came from seeing how much better pilots were on the last day of the exercise as opposed to when they first arrived.

3

Preface

In first person story telling it is important that the story, not the author is the main focus, letting the narrative sketch the story for the reader.

Red Flag does something no other book has done before; it takes readers into the dusty, isolated world of the threat range, Range Ops, and to Canada. It lets the reader see what this life was like and what made the ground threat tick.

I will also re-live the development and installation of an entirely new digital communications net and the start of creating the Red Flag Measurement and Debriefing System (RFMDS) and follow-on Individual Combat Aircrew Display System (ICADS) and Nellis Air Combat Training System (NACTS) that has been praised as one of the most important Red Flag enhancements since the operation began.

It's a unique world populated by unique people doing special things, a place unlike anywhere else. And you, the reader, for the first time, get to share it with them.

With a few exceptions, I do not use names; partly because it would not be fair without an individual's permission, and tracking them all down would be impossible. Names are not critical to the story.

In addition to explaining what is going on, I will take the reader with me on day-to-day range operations. As much as possible, this will include the sights, sounds, smell, and feel of the ground threat range in operation. While I may not be able to cover everything, I will share as much as possible.

Kernan Chaisson, Capt. USAF, (Ret)

Introduction

R ed Flag, the ultimate air combat experience, is described with all sorts of definitives: the definitive flying experience, the definitive air combat challenge, the definitive maintenance experience. There is one other ultimate part; the ground threat range, that always respected and usually surprising ground threat range. While success against other fighters and getting bombs on target brings bragging rights, success against the advanced threat simulators in the hands of extraordinarily skilled airmen and civilians brings survival in the real world of combat.

The world has changed and American fliers are less and less likely to face a peer-level air-to-air force. The challenge from the ground has become the biggest threat fliers are likely to face. Learning to counter these threats is critically important.

Red Flag originated as a way to give pilots combat experience against an Aggressor Air Force which simulates the aircraft and tactics they were likely to face should they have to fight. But over the years another part of Red Flag developed and grew into a crucial part of the experience, the threat from the ground. The extensive threat range complex became one of the most valuable parts of Red Flag.

There have been movies glamorizing fighter pilots and their air-to-air combat training experience (The '80s film *Red Flag: The Ultimate Game* was short-lived, fun to make, but except for the action shots pretty so-so as movies go).

Much better was the 2005 release IMAX *Fighter Pilot: Operation Red Flag*. It was released and well received at the AOC (Association of Old Crows) International convention.

There has been a lot of attention to the fighter pilots and their jets; but precious little about life on the ground at the all-important threat range. Little has showed what Red Flag combat is like looking up, firing the simulated guns and missiles at the airplanes trying to accomplish their missions, and working with the pilots on developing skills that will protect them from the killers on the ground. The threat crews tend to be invisible; a buzz from the threat warner, voice on a video tape, a score on a chart. Isolated on the range's desert floor, they are not seen or known. But what they do is important, vital, life-saving.

A lot goes into making the ground threat work; creating the environment that turned Red Flag into the premiere air combat training experience that it is. This book presents the experience I had as Red Flag Chief of Threat Analysis in the early 1980s running the threat range and debriefing the aircrews.

This was followed by a tour as Range Group Communications Project Engineer, developing and installing a new, advanced digital microwave system for the range, helping create what would make possible the computer-based Red Flag Measurement and Debriefing System (RFMDS) that took advantage of new technology to bring range threat control and crew debriefing into the new age.

Air-to-air combat training has become less critical as the world's military environment changes and the United States faces fewer peer competitors in the air. Allied air forces can expect near-absolute aerial supremacy; but the ground-to-air threat has spread and exponentially increased in deadliness. This makes the development and updating of EW threat ranges of the utmost importance. It is not skill in dogfighting but the ability to avoid small missiles, gunfire, and similar weapons that saves lives today.

The Nellis range can be re-configured to represent what pilots should expect. A major effort to re-build the arena for pilots to train for combat in the Persian Gulf was the beginning of an on-going program. It has been kept as current as possible for later combat in Iraq and Afghanistan.

This is not always easy when small arms and rocket-propelled grenades make up the bulk of the threat; but pilots can still be given an opportunity to practice ground attack while incorporating maneuvers designed to reduce the likelihood of being hit from the ground.

There is also the opportunity to practice some of the electronic countermeasures that can negate the ability of enemy forces to communicate. The range area has been used for testing and tactics development for unique operations that continued with the expansion into cyber warfare and special operations.

The late '80s and early '90s was an interesting time. Computers were coming into more general use; but were bulky, slow, and memory-starved by today's standards. Displays tended to be bland alphanumerics and there were no cell phones. "Text me" was not yet in the vocabulary and email was rudimentary at best.

In a March, 2013, *Washington Post* feature article on the F-35 program, Lt. Gen Frank Garner, Assistant Vice Chief of Staff, spoke of "the way American's go to war – we don't want to win 51 – 49. We want to win 99 – 0." Red Flag was designed to hone the skills aircrews need to insure this advantage.

This is not a diary or taken from contemporaneous notes. It is the story of my tour at Red Flag based on what turned out to be a surprisingly extensive visual and emotional memory. I was surprised at how many things came back with clarity and in detail. *Red Flag* does not include everything, but it covers a lot.

Because a range operation such as Red Flag involves classified information, producing a book such as this can be tricky. Fortunately, most of the restricted information involved technical material (frequencies, power levels, tactics, etc.), that would add little to the story. Because of a familiarity with the rules, I made sure no restricted information found its way into the book. In the twenty-plus years since I was involved, many of these things found their way into the open, unclassified world.

Some stuff did not come back and some stories still cannot or should not be told. But there is a much that can be shared.

Introduction

It seems that all good Air Force pilot stories begin "There I was, at 30,000 feet ...." But this is not a pilot story. It is the story of a bunch of non-pilots helping pilots learn an important part of their jobs – survival. So here it is.

"There I was, out on the desert floor ...."

* * * *

1. What Is Red Flag?

Why and Where?

From 1965 to 1973, combat pilots in Viet Nam were not faring well in air-to-air combat over the north. An Air Force analysis, *Project Red Baron*, found that once a pilot had gotten ten combat missions under his belt survival chances increased dramatically.

The (then) commander of the Tactical Air Command, General Robert J. Dixon, accepted a concept created by Colonel Richard "Moody" Suter that would give pilots a chance to fly these first ten missions in a safe and productive learning environment. Col Suter came up with his concept after talking to pilots and crews world-wide. The 4440th Tactical Fighter Training Group (Red Flag) was chartered 1 March 1975 at Nellis AFB, just north of Las Vegas, Nevada.

Las Vegas went from hot, hard desert to lush, wild gambling oasis in a made-for-movie story of booze, sex, and gambling. The area to the northwest was unique – one of the only places on earth on which officials decided to drop atomic bombs. In January, 1941, the Army Air Corps established a rustic airfield that Micheal Skinner (in *Red Flag: Air Combat for the '80s*, p.52) described as a dirt runway, water well, and dinky operations shack. The weather was great for flying (severe clear most of the time) and the airfield was on the edge of an enormous expanse of government land. The land was cheap and no commercial interest was going to fight the Army for it. So a natural use turned out to be gunnery practice. According to Skinner, at one point Las Vegas Army Air Field was turning out 800 B-17 crewmen every five weeks.

The field was shut down in 1947, re-opening as Las Vegas Air Force Base in 1949. It was re-named in honor of a P-47 pilot from Clark County, Lt. William Nellis, who was killed in Europe in 1944.

Over the years, the base and its mission grew, as did the areas that would come to be the Nellis Range Complex. Not just the tactical air force, but the Department of Energy and several other powerful, secretive government agencies set up housekeeping in the area north of the base. This included the birth of aircraft such as the U-2 and SR-71 in a square box hidden away in the mountains at the mysterious *Area 51* that everyone knew about but no one could officially talk about. Over the years, *Dreamland,* as it came to be called around Nellis, became a key research facility for top secret aviation and electronics projects of all sorts. In addition to advanced aircraft, *Area 51* gave birth to some of the wackiest lore imaginable; aliens in cold storage, hundred-mile-long tunnels to where and for what purpose no one knows, flying saucers (well this one was not all that far off), and annual gatherings of UFO buffs and other whack-jobs along the road on the east side of the range.

The nation's nuclear test site is located between the main base and the Tolicha Peak range complex, just under a hundred miles up a public highway. Preparations for the moon landing used some of the testing complex further up range, as did the team that developed the Mars rover vehicles. There would even be a complex adjoining the northern electronic warfare range that was even more hush-hush than *Area 51.* That is where the nation's first stealth aircraft was developed and tested in absolute secrecy.

The Tonopah Test Range was even more secret than *Area 51* and came to be called by some *Area 52.* (*Area 51*, pg. 342) Operating from a corner of one of the most active ranges anywhere, keeping what went on there secret from even an accidental glimpse could be a challenge. It is some serious classification when the final security out-brief lasts two days.

Nellis became home to the 57th Fighter Weapons School, the hardest and most prestigious training a fighter pilot can receive;

where they train those who would train the rest of the force. It is the toughest air combat course there is and its graduates are the top of the heap, bar none. Nellis is also where the Thunderbirds live.

The Operation Is Born

Red Flag took advantage of 5,000 square miles of military-only flyable airspace and, 2.9 million acres of land within which to practice air-to-air combat with *Aggressor* aircraft flying enemy tactics. There is another 7,000 square miles of MOA (Military Operating Area) airspace shared with civilian aircraft. This is the largest area where such training and experimentation can take place, to include the use of live weapons.

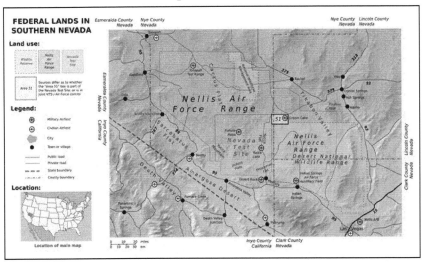

Nellis Range Area (Source: State of Nevada)

Keith J. Masbach, President of the Geo-spatial Intelligence Foundation, speaking at a joint International Spy Museum/Smithsonian Resident Associates course in 2011 spoke to what Nellis has become today. "From the mid-80s, the government had been building an area in the California and Nevada deserts to practice for the Cold War. Ft. Irwin and the Nellis AFB Ranges were initially independent but eventually interconnected by a digital microwave (NDMCS) link through a pass west of the Nellis range." It was a shot that was tight but doable. I can vouch for this because my digital microwave survey team found it, although it would be years before the Air Force would actually install the link. "There were questions of the value of the system when the USSR collapsed. But Saddam Hussein 'came to the rescue'. Operation Desert Storm used a reconfigured range to develop many of the tactics that resulted in a quick but effective dismantling of the Iraqi force."

A Good Plan Comes Together

Red Flag developed, and continues to use a proven plan where squadrons deploy to Nellis just as they would to a combat arena. Typical Red Flags were conducted four times a year, with another Maple Flag operation lasting four weeks in northern Alberta, Canada. A squadron would normally participate for two weeks, with a different squadron taking part the second two weeks.

The scenario planned for the second section was usually a repeat the first two weeks. The idea was for crews to fly in, become familiar with the range and surrounding area, then work gradually through an increasingly hard 'war' until the "Big Gorilla" on the last day. While in the early days of the exercise missions may be spread out among the participants. The 'Big Gorilla' puts everyone up at the same time in a mass battle that tests all of the skills practiced through the preceding weeks.

It can be a thrill on that last day to see and hear the mass takeoff where everyone heads of to what is usually planned by the Red Flag staff to be one helluva fight and a chance for the crews to really test themselves. By this time landmarks like Gold Flat, Kawich Valley, Cedar Peak and Cedar Pass, as well as Quartzite Mountain, Belted Peak, and Mount Helen are thoroughly familiar to everyone. Everyone left had only seen Bald Mountain from the north side, not having "busted the box" and getting into very deep doo-doo with *Dreamland* security.

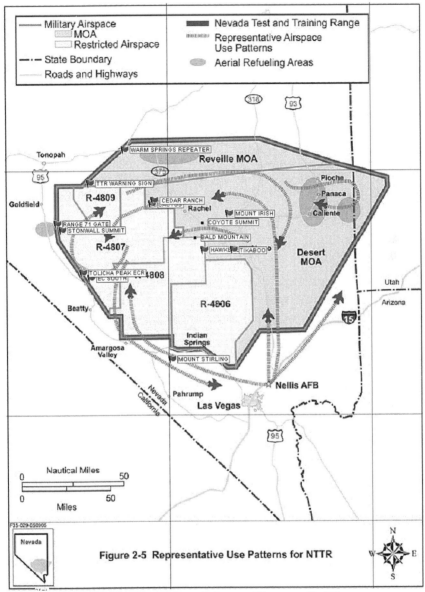

Figure 2-5 Representative Use Patterns for NTTR

Nellis Operational Area (Source: DoD)

What Is Red Flag?

The Red Flag scenario is rather simple. The bad guys, for some reason, decide to attack Las Vegas. Their battle lines are arrayed so they attack to the East (even though Las Vegas is actually to the South). There is a FEBA (Forward Edge of the Battlefield Area) just after Coyote and Cedar Pass, the range entry point. It is there the *Aggressors* lay in wait to attack, disrupt, and in general make life miserable for the attackers. Actually, the exercise scenarios make up names for the country in dispute; but we sometimes called it Vegas just for the hell of it.

The attackers bring their own Combat Air Patrol (CAP) to defend against the full-time hostile air force. Sometimes out-of-town participants, particularly Navy F-14s in those days, joined the defense force to practice their air-to-air skills. Off to the east would be an AWACS to try and control the attack, as well as tankers to refuel the force. Range Control, depending on the scenario, could call a 'mort' on attackers determined to have been shot by the defenders; at which point they had to fly back to the entry and be reconstituted into a new attacker, where they turn west and try again.

Further into the range area the attackers would find an array of targets, including an industrial complex, convoys, and a couple of airfields ready-made for deep interdiction. Just to be sure no one had any excess fun, once they got into the range's full-fight area, surface-to-air missile and gun simulators were spread around to further complicate the attackers' lives. They were especially thick around the airfield and industrial complex targets and gave the participants a lively set of SA-2, -3, and -6 SAM threats to dodge and neutralize with jammers. ZSU-23-4 and 57mm anti-aircraft guns were tossed in for good measure. The SA-7 shoulder-mounted SAM was also available to be deployed on select missions.

Clever engineering created signals that caused the attackers' sensors to see what appeared to be the real thing. Great care went into antenna transmit patterns so the received signals appeared authentic to warning receivers. The development of a small cardboard and Styrofoam rocket added a visual component that was so startling that

Smokey SAMs were said to be responsible for an up-tick in laundry bills back at base. More about this later.

Although most of the simulators were immobile by nature there was a particularly nasty little devil that could rove the range. The M-114 lightly armored tactical vehicle that never caught on with the Army and the radar from an F-5 were not the most sophisticated systems ever. But in the hands of the 554th Range Group, they become a particularly deadly anti-aircraft gun, the ZSU-23/4 (without bullets). Participants never knew where they would show up, just like in real combat. The whole idea was for the fight to be as realistic as possible.

M-114 Visimod. A ZSU-23 Look-Alike. (Source: U.S. Army)

The combat progressed as it would in a real Cold War attack in Europe. Every step of the way Range Group and Red Flag staff recorded as much as possible to use in post-mission de-briefs back at RFHQ. There were some targets where attackers could use live munitions on select missions. It sometimes got tricky to make sure the attack did not happen where range people were on the ground.

Targets up range were, in some cases, very realistic. Thanks to painted telephone poles, wrecked vehicles, plywood and canvas; pilots faced a realistic target set impossible to create at bombing ranges they worked back at the home-drome.

The targets appeared dangerously realistic in one case. One day, during the time in between Red Flags when Fighter Weapons School was the only range user, a private pilot developed some kind of mechanical trouble while flying in the Tonopah area. He managed to stray across the range complex border. His pleas over the emergency frequency said he just had to land and was going to use the airstrip he saw just off the nose. He felt he had no choice; he was in trouble.

The weapons controllers came up on the radio and convinced him to slide his Cessna over to the Tonopah airport a few miles to the southwest. Whatever his problem was, it would be no match to what he would have faced had he actually put down on the airstrip he spotted. A four-ship of attackers was boring in on the target loaded wall-to-wall with live munitions.

A Day in the Life Of.......

A day involved mission planning by the players starting early in the morning. About mid-morning, the "morning go" took off for what usually took a couple of hours from first takeoff to final landing. Informal de-briefs, lunch, and preparing for an afternoon mission took up the middle of the day. In mid-afternoon the players did it again.

Late in the afternoon everyone gathered in the main auditorium where the official scoring and evaluation of the day's activities took place. This is also where squadrons explained their tactical plan to everyone else. Plaques from units who had participated in previous Red Flags looked down on the tired, sweat-stained, and anxious flyers.

The *Aggressors*, target scorers, operations staff, and threat analysts used gun camera film, scoring forms, and treat video to let the group know how they did. There was belly-aching, hoots, cheers (for particularly clever actions by a crew) and information important for the next day's missions. An orange flag with the word "bulls**t" sometimes would wave from the audience as a pilot called attention to a call he did not agree with – usually to no avail.

The F-111D (Photo: U.S. Air Force)

A typical Red Flag saw the involvement of just about every aircraft in the inventory. F-15s were major CAP players. Allies often provided air-to-air fighters as well. F-4s, F-16s, F/A-18s, F-111s, A-6s, A-7s, A-10s, CH-53s, C-130s, B-52s, just about everything in the inventory of the time could be involved. British Jaguars and Buccaneers were regular participants, as were the Australian F-111Cs. The French came with Jaguars and Saudi Arabian Air Force brought their souped-up F-5s. Allied participation would increase over the years with some unusual pairings on the Red Flag ramp.

Way Up North

Once a year, Red Flag packed up and headed north to Canadian Forces Base Cold Lake in northern Alberta. The large base had lots of space for visitors and their airplanes and a massive wilderness over which to play. This made it an ideal place for assertive tactics and participation from a large number of international players. British and

Marine AV-8B Harriers were regular participants, as were the German *Luftwaffe*. The Canadian home boys took part with their F-5s and F-18s. The base helicopter detachment was invaluable in supporting the exercise and as the rescue force, both exercise and real.

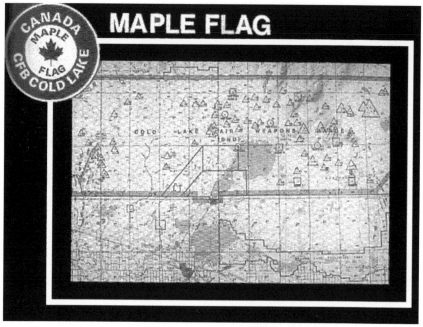

(Source: Canadian Forces CFB Cold Lake)

****

2. The Base Called Nellis

Nellis Air Force Base, Nevada, is known world-wide as "The Home of the Fighter Pilot. Located on the Northern edge of Las Vegas, it has been associated with airplanes, weapons, and tactics.

The Front Gate. (Photo: U.S. Air Force)

In case you didn't drive. (Photo: U.S. Air Force)

19

The Base Called Nellis

Dawn hitting Sunrise mountain on the north edge of the base turned it red and gold. The colors changed by the minute. A rare, light dusting of snow added to the wonder of a new day as it was being greeted by the song of jet engines running up and the smell of JP-5 on a crisp morning.

Sunrise Mountain (Photo: U.S. Air Force)

Quiet times on the ramp and runways would sometimes be split by the roar of engines in the run-up cells at the end of the field. The nose-tickling, sweet-burned aroma of flight line activities gave the feel of power moving around the ramp.

The ramp always seemed vacant and sleepy when only local jets were there. It filled up and came alive during a Red Flag. It was the same with the skies. The ramp again became lonesome and forlorn when participants departed.

Nellis was never a silent place. Fighter Weapons School, the Thunderbirds, and others made it the top of the fighter world mountain. But an active Red Flag brought a special life and vitality to the place. A palpable excitement permeated everything, with the morning and afternoon takeoffs adding a special zing to the atmosphere. There was an almost visceral thrill as afterburners shook the world and fighter jets streamed off to the north.

Nellis Ramp (Photo: U.S. Air Force)

Aggressor F-16 Taking Off. (Photo: U.S. Air Force)

It was especially thrilling when the final massed flights topped off the two weeks. It was a way of proving what had been gained during the intense exercises.

Two decades later, while visiting Las Vegas, I saw the joint takeoff of several fighters, a B-2, and an F-117. It gave me chills.

B-2 and F-117s in Formation (Photo: U.S. Air Force)

* * *

3. How I Got There

Confessions of a draft dodger, 1961.

I had no particular feelings one way or another about the Viet Nam War that was going on. "What do you want to be in?" asked the Air Force recruiter. "Not the Army," answered I, who had just found out that he had received a very low draft number. I preferred a dry bed and cold beer to a wet bed and cold food. The Air Force would probably do the trick.

"I have a slot in electronics, radar to be exact," he informed me. "Any interest?"

"Sounds good to me." Having no idea what that meant. I just knew it would probably be better than the Infantry.

The rest is history. I set out on a career where the Cold War would be descriptive, not just a name. It was a career filled with adventures and opportunities that made the next 23 years, 7 months, 30 days a time when, as one compatriot put it "they even pay me to have so much fun!" It may not always have been fun; but it was always a time of new things, opportunities, rewards, and unique experiences.

School Time

After that wonderfully character-building exercise called Basic Training, the Air Force sent me to the sun-drenched, white sand beaches of Biloxi, Mississippi; Keesler AFB, to be exact. The Electronics Training Center of the Air Force was, and still is, where airmen, NCOs, and officers went to learn to maintain and operate the various electronic and communications systems in the service's inventory.

How I Got There

Keesler AFB, Main Gate. (Photo: Keesler AFB, Mississippi)

Electronics school in the early '60s approached what could be thought of as close to a basic engineering course today. This was the time where actual resistors, capacitors, coils, and vacuum tubes made up circuits. Troubleshooting often involved calculating voltage and current as well as measuring the values at various points to decide if a component needed replacing; manually unsoldering it, and putting in a new one. Keeping a system running involved knowing a lot of electronics theory.

For me, it was learning to maintain the heavy-duty, long range AC&W (Aircraft Control and Warning) radars that formed the Distant Early Warning (DEW) Line stretching around the nation's borders and forming a multi-layer detection fence across the top of the North American continent to protect the United States and Canada from attack by Soviet bombers.

What did not yet exist was the digital world of multiple circuits on a single silicon chip. We had no idea of today's semiconductor chips with hundreds of circuits in a small silicon wafer and today's engineers probably get just a little taste of our vacuum tube world in a course on electronics history.

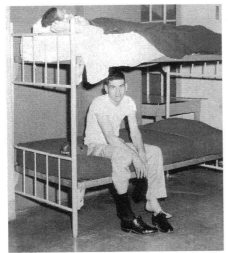

Everyday a picnic, every meal a feast,
(Photo: Keesler AFB, Mississippi)

A class in basic circuits. (Photo: Keesler AFB Mississippi)

Night and correspondence courses abounded, such as CREI, the Capitol Radio Engineering Institute. Radio Shack carried kits to build radios, TVs, testers, and the like. Certificate courses in radio engineering, etc., did not go as far as advanced calculus; but a graduate knew things not touched on in today's modular, digital world. Digital was not common anywhere. Labs were, however, investigating and developing this for future hardware.

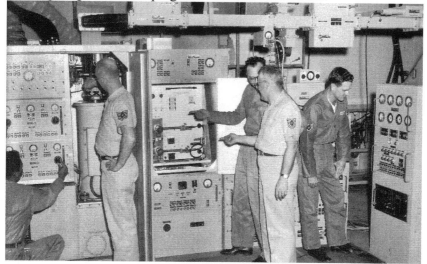

Radar class (Photo: Keesler AFB Mississippi)

Troubleshooting involved many of the same calculations as the design effort. There were no hand-held calculators or computers as today, and most work beyond simple addition and subtraction was done with a slide rule. How many of today's engineers remember those? My brother and I designed a case containing a slide rule to mount on the wall of a computer room or engineering office. It was marked "in case of emergency, break glass."

We also came up with another interesting send up. It would be fun to walk around a campus or tech training center today with a slide rule in its case suspended from your belt (at least a 12" or 15") and watch the puzzled looks at what students used in the days before calculators. It was always fun to explain that these things were used to design early electronics and aircraft; and to get the first rockets into space. Computers were just beginning to find their way into the real world and were not readily available to individuals. They were rudimentary devices at best, not what we have today.

How many of us today can even balance our checkbooks without a calculator? My first four-function (add, subtract, multiply, divide) bought in the '60s cost over $100 and a set of batteries lasted a couple of hours. Just the other day I saw a full scientific graphing calculator . for sale at the supermarket.

Another interesting experience was learning computer programming using punch cards (remember those?); with all their little holes. Any one out of place and the whole process would stop dead in its tracks, giving no clue as to what was wrong.

Technicians and engineers usually had distinctive calluses on their thumb and index fingers, the result of pulling hot vacuum tubes from their sockets and putting new ones in to get a system back on the air. Those were the days when if one's TV set quit, you took all the tubes down to the local drug store (pharmacy in today's language) and checked them in a tube tester, replacing any that showed weak. This made young airman radar or radio technicians very popular at family gatherings. Believe it or not, one had to get up and actually walk to the set to turn it on, adjust the volume, or change channels – among the four that existed then.

Notice I said nothing about computer technician or IT specialist. Those jobs would be another decade in the making.

A Career Begins.

Graduation from Tech School was followed by driving to my first real job as a radar mechanic – two blocks down the street from school to the maintenance squadron, not realizing how valuable the next few years would be in developing my knowledge and skill with radars and electronics. This was the first step in what, looking back, was a path that would begin my setup for a future Red Flag job.

Complete radar sets provide students hands-on experience.
(Photo: Keesler AFB, Mississippi)

The Keesler job was to maintain the radars used in tech school classes, the ones I'd just come through as a student. Most graduates go to a radar site where they spend their time on one or two types of radars that were designed for reliability, usually needing little more than regular alignment and scheduled preventive maintenance. When there was a major failure, the big guns took over and the newbies mostly watched or acted as gofers. Skills development can be slow in that scenario.

At Keesler, we were responsible for one or two of every long-range radar in the Air Force inventory. The job was to have them operational for class so the students could get direct, hands-on experience with the equipment and perform some of the maintenance tasks they would be called on to do once in the field.

With a constant parade of inexperienced, ham-fisted young airmen messing with the gear, that wasn't easy. It was also necessary, after each class shift, to find and repair the sample troubles instructors put into the system for the troubleshooting phase of training, troubles they often forgot to remove when class was over. Tubes with cut socket pins or adjustments out of limits were common. I would see, troubleshoot, and repair more problems in a shift than most line maintainers would see in a year.

Tech School radar training site.
(Photo: Keesler AFB, Mississippi)

I had some interesting experiences while in the maintenance group. Mother Nature decided that the Gulf Coast was a great place to hammer with hurricanes on a regular basis. One time weather forecasters predicted a direct hit on the Biloxi area. It was decided that to prevent damage, a couple of radar antennas should be taken down from their towers outside the hangers used for radar school. I got the task of climbing up and disconnecting the sail on one so a crane could set it down on the ground.

Naturally, by the time the crane got to the AN/FPS-8, the first wind and rain bands started to hit. I scampered up the tower and began loosening the mounting bolts. My head was extended up through the bottom antenna braces and turntable. Not smart, but there I was. Just as I loosened the last bolt, a gust hit, yanking the antenna sail from its mount. I was able to duck just in time to keep from suddenly becoming about ten inches shorter. The bottom brace snagged my fatigue hat, so the main damage, in spite of everything, was that I had to go buy a new one. That pissed me off. It also taught me to never climb an antenna while suffering from the effects of cranio-rectal insertion.

Although most of us in maintenance had not logged much more Air Force time than students, we were permanent party and real maintenance technicians; so the students trusted us to know more and act better than anyone else. Was that ever a mistake.

The old hanger housed full-sized radars for students to learn maintenance procedures. They transmitted their radio frequency power into a dummy load instead of an antenna. For a while, however, there was an old set which had both a dummy load and antenna that would radiate radio-frequency energy with the flip of a switch.

It was common in those days for 'volunteer' airmen to do cleanup duty. One project was to haul burned out fluorescent tubes to the dumpster. The sight of a young airmen walking through the hanger with their arms full of these tubes was just too much temptation to pass up. We'd turn on the transmitter, switch in the antenna, and swing the beam over our intended victim. Although the lights no longer worked, the radio frequency energy from the radar energized the gas in the tubes and they lit up. This was disconcerting to say the least, and an armload of neon lights got thrown into the air or onto the ground. The instructor supervisors were not real happy to have their students treated that way; which meant we had to be careful not to get caught. A permanent solution came when the offending radar was dropped from the curriculum and hauled away.

A major project was the installation of a new AN/FPS-27 search radar. It was the most powerful radar (5 million Watts) of its time and had all of the latest receiver techniques. It took a five-story building to hold it; so everything came in the size Extra Large. One day after the system had been in use for just under a year, the transmitter pulse transformer arced internally and quit working. This disabled the transmitter and crippled the training course. It would be at least eight months before the factory could build and send a replacement, and that would cost over $1.5 million.

A team of us drained the insulating oil from the damaged pulse transformer, about 10' by 12' by 8', with a small hatch on top, and crawled inside to investigate. Although most thought it could not be done, we decided to take a stab at fixing the transformer's damaged wiring, wires the size of water pipes. We managed to repair them, but the tricky part was making sure that all contamination was eliminated. We went after the job with enthusiasm, dedication, and buckets of trichloroethylene. We'd been using the stuff to clean everything. It was great on grimy circuits (with actual resistors, capacitors, and coils); so why not here? It cleaned things instantly then evaporated, leaving a shiny clean surface behind.

This was before trichlor was declared extremely hazardous and banned by the Air Force and just about everywhere else in the world. We proved how dizzy and daffy one could get crawling around inside a big box, wiping the stuff all over the sides to get rid of unwanted substances that could contaminate the new insulating oil and cause more arching when the radar was fired up. We were overjoyed when, repair done, the transmit button was pushed for the first time and there was a loud buzz, not a big bang. The Air Force recognized the job as one of the top money-saving, creative efforts of that year.

I participated in another stunt for which I should not be proud. In those days the radar display scopes used cathode ray tube technology and rotating coils to create the images operators looked it, getting these signals to the rotating coils by way of metal slip rings. Thus the evil genius reared its head. We maintenance wizards knew that the students watched everything we did so they could repeat it when they got on the job. When the front or top of a scope was opened to access the innards and there was a group of students doing a class project on a nearby scope, we would lean across the cabinet, extend the lead from one of those old GI automatic pencils, and sweep the lead across the slip rings. The result was a lovely display of colorful sparks. The voltage was very high, the amperage very low, so there was little risk of electrocution.

The students were really impressed and wanted to be able to do that themselves. So first chance they got, the copy-cats tried it, getting knocked on their butts as a result, wondering why we could do it and they couldn't. We had a secret the gullible students were not aware of. Those pencils had a plastic body and metal pocket clip and eraser holder. The trick was to hold the pencil and touch only the plastic, a trick they did not know.

Luckily, no one was ever hurt. We quickly were nowhere to be found and thus could not be blamed for what happened. The instructors knew damn good and well who the guilty parties were that 'electrocuted' their students; but with no witnesses what could they do, except turn it into a learning experience for their classes, warning them about how careful one had to be around high voltage, especially when they do tasks anywhere around signs that say "Danger, High Voltage"

I also learned some of the vagaries of the Air force supply system while in the maintenance group. Once we needed some metal screws for a job and ordered a dozen. Unfortunately, a coding mistake at supply ordered a dozen gross cases (which each contained 24 boxes, which each contained 100 packs, each with 50 screws). For all I know, with backlogs and such, those damn screws might still be coming in to Keesler.

In another supply system glitch, we needed a 5,000 lb winch. A typo had all of us anxiously waiting for the arrival of the 5,000 lb wench the supply form called for.

The Real Cold War

After three interesting years, it came time for my year as a true Cold Warrior, a tour at a DEW Line site in remote Alaska. The 717th AC&W Squadron was in the middle of the state. Tatalina Air Force Station was built on Takotna Mountain and adjacent to the Village of Takotna, which was located at the base of Tatalina Mountain. Yep, it would be that kind of year. It would be a time for learning to deal with boredom, getting along with 18 guys on a mountaintop isolated from the remote site. This was my first experience working on radars operating in the real world, watching the skies for Soviet bombers. One learns a lot about tuning systems for peak performance, and about fixing problems quickly, knowing no one was coming to help. We were on our own 24/7.

Top Camp (Photo: Radomes.com)

33

How I Got There

It was quite an adventure, that obligatory year of frozen isolation and constant alert. The hilltop site is where the search and height-finder radars were located, along with ground-to-air radios and the Western Electric White Alice tropospheric scatter long-haul communications site. Sleeping quarters, recreation room, kitchen, bar; it was all there on the peak. Everything we used, food, water, supplies, beer; had to be hauled up a twisting, narrow road clawed out of the mountainside. At one point, we were closed in totally for over a week by a blizzard and white-out conditions. Nothing could come up, including water. Our cook proved how many different ways fish sticks and beans could be prepared.

When the weather broke, our cat-skinner immediately began clearing the road up the mountain so supplies and the water truck could make it up. We were down to two inches of water in our storage tank. The plowing would take some time, so the Commander hopped into a Sno-Cat and came up, bringing a couple bags of mail and several bottles of whiskey for everything we'd been through.

 Two USO performers, a musician and a contortionist, who were part of a small troupe making the rounds and entertaining the troops at remote sites. These two decided to come up to Hilltop and put on a show for us. They ended up stranded with us. The USO gets praised for visiting the troops and putting on shows at big bases overseas. But it's the little troupes like this that are the real heroes – coming to godforsaken sites to bring a little entertainment to a handful of GIs stuck there.

A husband and wife team that specialized in a unique performance where she put a black bag over his head and he very quickly painted small snow-capped mountain scenes while blindfolded. They then gave the paintings to members of the audience, a unique keepsake. After putting on their show at a sister site, they were killed in an airplane crash. These small USO Tours never got the credit they deserve.

A Bracing Day Up Top   (Photo: AFRSV.com)

Moose, bears, wolves, and mosquito swarms that would show up on the radar scopes; these were our neighbors. In the summer, a couple from Miami ran a successful gold claim in the valley below the site. Site people were always welcome and would usually bring some beer, so they would let us pan a stream that was rich in ore but too expensive to work commercially. Several guys panned some gold dust to take home. I scored a nice little nugget.

The Rosanders had rescued a baby beaver when its dam collapsed. They kept it in their cabin. He was a cute little character, building a 'dam' in front of the door every night, using shoes, clothes,

35

anything he would find on the floor. He really liked it when someone would hold him in their arm like a baby, and feed him pancakes. This was a favorite activity during our visits.

One especially wacky adventure came when our Senior Master Sargent maintenance chief almost conned *ABC's Wide World of Sports* into coming up to televise the annual Midnight Sun Softball Game that pitted the site against the Takotna locals in a game of beer ball that started at midnight the day the sun never went below the horizon. (For those not in the know, beer ball is played with a can of beer placed on each base. A runner cannot advance past that base until he has chugged the whole beer. Things can get very interesting). Before the project got too expensive and it was too late to turn back, someone at ABC had a sudden attack of common sense and sent a politely-worded letter informing us that the network had decided to pursue other programming on that day.

In those days a Cold Warrior made two trips, one in and one out to and from the site. If the **out** trip was less than a year from the **in** trip you were probably either going to the hospital or dead. It would be years before leave from a remote site was allowed. Eventually the FAA took over some sites. Others were replaced with unmanned radars that were operated remotely and got by with only occasional visits by maintenance technicians who made sure everything was up to snuff.

Life at these remote stations could be lonely and boring; but also interesting and memorable, providing stories that would last a lifetime. Before the snow closed in, site personnel had the opportunity to hike the tundra and visit remnants of the gold mining days. Not far from Tatalina was the old, abandoned camp of Ophir. Many of the buildings still had canned goods and other stuff on shelves as they were when the old sourdough miners left. There was also the remains of the shack where the 'girls of the evening' relieved the '49rs of much of their gold. An old dredge was sunk in the sluice pond, a scene directly from a Robert W. Service poem. Believe it or not, I found that a company is back in there mining gold commercially.

Moving around on the tundra caused a black cloud of blood-thirsty mosquitoes to rise from the permafrost. They had little time to find warm blood, and they did not care where it came from, but wandering Air Force personnel were a favorite. A beautiful, massive wolf and her cubs were frequent visitors to a clearing below the back side of the mountain, and the weather forecasters at the landing strip had to routinely chase moose off the runway so the twice-weekly Northern Consolidated F-27 flight could land.

The site also maintained a fish camp on the Kuskokwim River where small groups could take a little time and boat down there not so much to fish as to not be at the site for a couple of days.

Some close relationships formed at the site. A cute little black and white dog named Sally grew very attached to one of the radio maintenance techs. If he went down hill on routine business all year, she thought nothing of it. But for some reason when he rotated out at the end of his tour she knew and had a heart attack running down the hill after the truck.

Bears will be bears and the black bears living around the site were not particularly aggressive or vicious; but they could be a real pain in the ass. They found that the site dump was like an Auto-Mat, the perfect place to grab a snack without working very hard to get it. Whenever anyone took at truck down to bottom camp, they took whatever garbage there was to the dump. Whenever a truck pulled up to the dump, the bears came a-running to see what we brought.

One day I made the run. Usually the bears were patient and waited for the truck to be unloaded before rummaging in the gifts being provided. But this one day I was pulling cans off when I heard noise from the other side of the truck. Looking up I saw this young bear on the other side, obviously impatient with the pace of my work, helping unload by pulling cans off himself. We spotted one another at the same time and each of us retreated away from the truck and stood there waiting to see who would make the first move. That turned out to be me, diving into the truck and taking off. Someone else could bring those three cans back up.

Speaking of bears, one day the weather was nice and the First Sergeant was sitting at a table close to an open window of the lower camp chow hall when he heard some noise and found a bear clawing through screen and trying to get to the NCO's lunch. Everybody shouting chased the bear away.

Lower camp in Summer (Photo: Radomes, Inc.)

A couple of days later at lower camp it was almost time for lunch when what should appear behind the serving line but a black bear walking along looking over the offerings. Probably the same one who tried snaring the First Shirt's lunch. Next to appear was this gangling, pimple-faced cook with a big soup ladle shouting and swinging. Luckily the bear remembered where the back door was and headed out. Unfortunately, the local game warden decided that this bear was getting a little too familiar with humans, could become dangerous, and had to be destroyed. A sad ending, but he was probably right.

One last bear story. At lower camp there were raised walkways between the buildings so people could go from one to another above the snow. It was summer, but I took a shortcut. Halfway across, a cute little roly-poly cub came bounding up to me wanting to play. I heard a grunt and there was Mama looking crankily at the two of us. "Hey kid, you're going to get us both in trouble." Luckily the little guy decided to go over and see what Mom wanted, and they both strolled off. I could let my breath out.

We had to make out own fun. There was a combination bar, rec room, NCO quarters in a building half-way between the main building and height finder tower. It is where any socializing that did not involve meals took place. We got our haircuts there from one of the guys who sort of could do a sort of decent crew cut.

It was also the location of the infamous (for Tatalina, at least) Johnson's Gap. The bathroom was located in the main building, quite a walk up a connecting (and cold) hallway. Ingenuity being what it is, we found that a door at the beginning of the heightfinder tower tunnel led to a sort-of flat spot and dropoff which came in handy, especially during movies while the reels were being changed. Oh, did I mention our bar and therefore beer was there? This proved the wisdom of the warning, "don't eat yellow snow." Somehow no one ever fell over, that we know of.

Being remote, everything that was done we had to do ourselves. This included fire fighter duty. There was one professional assigned who made sure fire hazards were eliminated and to take charge of any emergency response if, God forbid, it was ever necessary. Most of the protection was focused on lower camp. There was one Hilltop false alarm while I was there. The wind was howling at the time and it really scared the hell out of everyone. Truth of the matter was, given the way things were, if a fire ever broke out up top, we were pretty much screwed.

Speaking of having to do everything ourselves, if the Russians ever decided to come overland and attack, we had to defend ourselves. There were .45s for the weapons controllers and other officers at lower camp. There was a brace of M-1 carbines for us up top. We had to routinely exercise our best warrior skills; but at least the Air Force was smart enough to issue weapons but not to give us bullets. One thing we discovered; walking around the radar equipment with an M-1 slung over our shoulder was a sure-fire way to hit the cabinets and trip the safety interlocks (we kept the drawers open to be able to get into the system more quickly should something kick off). That invariably knocked the radar off the air. Finally, it was decided that even though we were required to be armed at all times, leaving our weapons at the maintenance desk rather that disabling the nation's northern defenses was the way to go.

The whole thing was not too bad in the summer if you had outside perimeter duty. The idea was, armed to the teeth and wearing our World War Two steel helmet with 717 AC&W painted on the front, we were to prevent any enemy incursion. Who the hell would ever want to invade out little piece of heaven was any body's guess; so the duty usually ended up a game of hide-and-seek among the 'guards'; with beer riding on the results of who caught who. In remote Alaska, you took fun from anywhere you could find it.

It is interesting how GIs everywhere are GIs, and there is an inclination to look out for one another whatever side they are on. The western radars in Alaska could pick up the Russian mail helicopter as it made the rounds from Russian radar site to Russian radar site every few days across the Bering Strait.

One time, the U.S. operators saw the echo disappear in between sites unexpectedly in a place it usually did not land. They then picked up a lot of aircraft obviously searching for something; but were looking in the wrong place, not where the helicopter disappeared from the radar screens. Pretty much on the QT, radio intercept operators who were Russian linguists and monitored the Russian aircraft frequencies quickly transmitted, in Russian, the location of where the mail helicopter apparently went down. There was a brief *spasibo* (Thanks) on the link and the search aircraft quickly converged on the location they had been given.

We had no way of knowing the results. To this day, I hope things worked out for the mail chopper crew. Even in the Cold War there were some warm spots.

Things in remote Alaska could be boring. The interceptor pilots knew that and sometimes tried to liven things up with what was called a 'bubble check'. The pilot would, without any notice, dive from altitude, level off on the deck, and make a high speed pass BETWEEN the two radomes. Sometimes the operators caught on just before the jet got to us. We'd try and run outside to watch an F-102 or -106 zoom by at nearly eye level. More often, there would be no warning and the no-notice *Boom!!!* of a nearly supersonic jet feet away. Scared the hell out of everyone. It took a little while to get our heart rates back to normal.

There were rumors that the Russian pilots, being pilots, would sometimes do a bubble check on our sites on the west coast, sites like Tin City and Kotzebue. That involved actually violating U.S. air space, so Anchorage scrambled jets to intercept the intruders. Actually, they had this down to a science where the Russians knew that they could wheel around and be back in their own air space before our guys got to them. Our guys knew that they would not reach them before they got back across the line.

There was a story that one time the Russians mis-calculated their turn and would have been caught by our interceptors had our pilots not eased off the throttle a bit to avoid getting into a position where they had to turn a prank into a war.

Heading South at Last

After a year at the pointy end of the air defense spear in Alaska, a scary bush pilot flight to the FAA town of McGrath, and a commercial flight to Anchorage got me to an "ol'shaky" C-124 "freedom bird" back to the Lower-48.

After a year at the site, seeing a long-in-the-tooth Piper Tri-Pacer show up to take me home was not as troubling as it probably should have been. The bush pilot loaded my duffel bag, many mail bags, and me in the back seat; with me on the bottom. A flight to the FAA town of McGrath was all that stood between me and civilization. But I had to get there.

There I was, jammed in the back of this old airplane when a strong cross wind came up. No biggie. We'd just wait for things to calm down and head out. Wrong! The bush pilot, greasy overalls and all, cranked the engine over, taxied part way down the gravel strip, and turned around. The runway crossed the leveled-off top of a hill that dropped off into a valley on all four sides. I should have been nervous; but I was going home. I could barely see from under the bags and the cross-wind should have been a worry. Bush pilots are noted for being crazy, but also for being damn good. He let 'er rip.

Running the trembling little plane to full throttle while standing on the brakes he waited for just the right moment. When it seemed the Piper was going to come apart at the seams, he released the brake and started a takeoff roll. We began to pick up speed and the end of the runway started coming closer. The pilot suddenly kicked the right rudder pedal real hard; stomped it would be a better description. It seemed like the plane jumped off the ground, turned into the wind, and dove off the side of the mountain into the valley. About half way to the valley floor we got enough airspeed to actually start flying toward McGrath.

No worries now. McGrath had a nice, real runway that handled commercial flights, one of which would carry me to Anchorage and the Lower 48. When we got there, it did not take long, there was this lovely runway running alongside the Kuskokwim River – with a gigantic construction crane right in the middle! After that takeoff from Tatalina how bad could this be? Little did I know.

The pilot started his approach, so I figured he'd buzz the crane and it would move so he could come around and land. Typical bush pilot, my guy had a different idea. Crossing over the crane he chopped the throttle. The plane dropped. Just above the ground he fire-walled the throttle and the little Tri-Pacer settled gently on the McGrath runway.

We taxied to the terminal building and unloaded all the stuff. As I slung the duffel over my shoulder, the pilot looked at the manifest and said, "humph, 375 pounds overweight," and walked off. I staggered into the snack bar. The one-armed waitress behind the counter took one look, "just came in with Walt, Eh?" and set two beers on the counter. She knew. What an end to my Alaska tour.

I was sent to a heart of the nation's air defense system, 20th Air Division Headquarters, Truax Field, Wisconsin. Truax hosted a SAGE (Semi-Automatic Ground Environment) blockhouse. It was one of 24 Direction Centers and three Combat Control Centers linked to the North American Air Defense Command (NORAD) Command and Control center via AT&T phone lines. This was a three-story, windowless square of concrete that housed the AN/FSQ-7 air defense computer, one of the first mainframe computers built by IBM. It drove computerized scopes that linked to similar systems around the nation to monitor aircraft in the division's airspace. Operators would direct air defense interceptors to ward off a bomber attack from the Soviet Union. The weapons controllers spent all of their time either practicing air defense or waiting for the computer to 'start over' (Ctrl-Alt-Del), something that happened with great regularity.

The FSQ-7 was an interesting beast. It was a massive, 250 ton computing machine taking up a half acre of floor space. The processor was built with 6,000 vacuum tube diodes that took up the entire wall of a room. The maintenance control office shared a wall with one of the processors. They glowed bright enough that we could read by them.

Memory 1 was 65,536 words while Memory 2 was 4,096 words, running on a combination of several oil-drum-sized spinning memory units and several 6-foot tall cabinets, IBM 762 magnetic tape drives to store data. The drum auxiliary memory had 50 fields of 2,048 words each. (Wikipedia, SAGE) This is interesting when viewed from a time where wrist watches use far more computer power just to tell what time it is.

How I Got There

***AN/FSQ-7 SAGE computer Main Frame and Control Center
(Photos: Wikipedia, SAGE)***

As a Maintenance Control Specialist, my job was to make sure that radar coverage was not compromised by radar sites in the division going off the air at the wrong time, like when the adjacent site was down for maintenance. That involved shuffling cards that reported the status, programmed maintenance or unscheduled repairs, and making sure the weapons control section knew if there were any blank spots in coverage.

We often ran simulated exercises where everyone practiced for a real-life nuclear attack. This invariably started late when we were hoping for a good night's sleep. But usually word got out ahead of time that a 'no-notice' exercise was planned. The exercise started with a simulated first detection and ran through an ever-increasing scenario. The DEFCON (Defense Condition) "Big Noise, Applejack, Delta"; meant nukes were imminent, and so was the end of the exercise and we'd be able to go home, unless you had day shift duty.

Such was life valiantly defending the homeland. I lived in the base trailer park, so at least all I had to do was run across a field to get to the block house and I could be crawling into bed before most of the other troopers had gotten off base.

Not my most stimulating job ever. For excitement, though, lose Air Force Two with the Vice President on board. One evening, Hubert Humphrey was traveling through our airspace, something he did routinely on Friday. His airplane crossed into our airspace just as I approved a site to go off the air for scheduled maintenance while the adjacent site was down for unscheduled repairs. This would not normally be a major problem; but with the VP coming through, the situation ratcheted up to federal-case level in a hurry. Air Force Two disappeared from the scopes, and the powers that be were not happy. It flowed downhill from there, stopping with you know who. About the only one who did not take a big chunk out of my rear that night was the Vice President himself.

NORAD tracking Santa

This was before the Internet. Today there is a major nation-wide program where Santa's journey is tracked on Christmas Eve. In those days, everything had to be done locally. At the SAGE blockhouse, we fielded calls from local kids on a special phone line. What fun trying to convince the excited tykes than Santa Clause was circling just outside the Madison city limits and they had better hit the bed and go to sleep as fast as they could.

What was that?

This was during the era of UFOs and the Project Blue Book craze. As with most paperwork no one wanted to do, Maintenance Control got stuck with it. While most of the sightings were easily explainable, there was one that never quite got figured out. The same lights in the sky were reported from a local girls school by the students, some airmen who were visiting their girlfriends, and the nuns who taught there. The Division Commander also reported seeing what had to be the same lights from another part of town. Let's see .... airmen, nuns, Division Commander, Hmmmm!?

My tour was put out of its misery by an Air Force decision to re-configure the Air Defense System, shutting down the 20th Air Division and routing the radar feeds to other SAGE centers.

The cooling system for a SAGE computer equipment had to be so powerful that once the computer was removed the Air Force sold the block house to the Oscar Mayer Company. The blockhouse was converted to cold storage for wieners and other meat products. No changes to the chillers were necessary to freeze anything put in there. Walls designed to protect from a nuclear attack easily kept the cold in.

SAGE Blockhouse (photo: www.coldwarpeacemuseum.org)

Up To Down East

From Wisconsin wieners it was off to the land of lobsters, Bucks Harbor, Maine. From being on the front line of airspace/area management it was back to the operational arena again. The 907th

Radar Squadron clung precariously to the rocky coast of Maine, "Down East", 28 miles from the U.S./Canadian border. It was truly the cold part of the Cold War. This included my car once being completely buried in snow for two weeks while parked in front of the site NCO Club.

Bucks Harbor and the surrounding villages could have been a scene from the movie *The Russians Are Coming! The Russians Are Coming!*, especially on those days when Soviet trawlers, sprouting all sorts of antennas, could be seen off shore. No telling who or what else was out there; but we had our suspicions. WMCS, the local radio station in neighboring Machias was located in the corner of a woman's living room and boasted that it was "next to the most powerful radio station in the world". This was true, but not quite the way it sounded. A few miles north was the Cutler Naval Radio Station which blasted signals to U.S. Submarines around the globe.

Not as remote as Alaska, The Bucks was far enough from regular supply channels and operational bases that we had to develop the ability to operate independently and creatively – a skill that would prove valuable at Red Flag and especially at Maple Flag years later. It was also a place where one could develop what could be a very expensive habit after moving 'down south' – delicious Maine Cold Water Lobster! The road from the site to Machias was lined with docks. It was routine to stop on the way home in the evening, toss a fisherman a buck or two, and drag some beauties right out of the trap. To this day my son complains about the trauma of riding in the back of my station wagon when I casually tossed supper in there with him. They did not have their claws rubber banded yet, so I guess he has a point.

Because the station was classified as semi-remote, the Mess Sergeant had the authority to buy food locally to save money. In that area one of the cheapest foods around most of the time was lobster, so he served a lot of it in the Mess Hall. Since visitors had to eat there, Bucks Harbor tended to be the most inspected site anywhere in the world. It was a rare week when we did not have some official visitor or other stopping by to "check things out", invariably in time for lunch.

It was also an encounter with the unusual. The AN/FPS-24 long-range search radar boasted an eighty-five ton antenna the size of a basketball court twirling away on top of a square five-story building. A person could actually do jumping jacks in the feed horn. This multi-megaWatt monster was said, jokingly, to be able to cook sea gulls as they flew by. The transmitted signal did cause the window screens in the NCO barracks at the bottom of the hill to "zing" every time the antenna swung past. Only twelve of these surface search systems were ever built. It was an interesting experience to work with this monster.

AN/FPS-24 Radar. (Photo: radomes.org)

My association with Project Blue Book continued. Since we were the only Air Force facility in the area, we had to handle the reports. They tended to be from various lobster fishermen and clam diggers on Saturday night. They usually came from around the same bar and the location reported was in the general direction of the Jones Port Light. Not too hard to figure those out. There was one case that took a little more investigation, especially since it reportedly involved an abduction. As it turned out, the abductees were a teenage boy and girl that suddenly went missing about the time there were reports of strange lights from a UFO on the ground in a field between the site and town.

A brief investigation found a burned-out plywood shed with two sets of clothes in the rubble. A couple of days later the two teens showed back up after hiding out at a friend's house. The story revealed that the UFO incident turned out to be they were almost caught in a "compromising situation" in the hunter shack. Scared half to death of getting caught, they departed swiftly, managing to set the shack on fire with their lamp in the process, leaving their clothes behind and ending up with their friend. Case closed.

One interesting experience says something about GIs and their special ability to find fun in almost any circumstance, especially where "adult supervision" is lacking. One weekend, a Soviet Naval ship made a port call at Saint John's, New Brunswick. Four of the crew figured they had enough time to realize a dream and go to a Stanley Cup playoff hockey game in Boston; a logistic challenge, but they were not about to let that stop them from trying. They were given the leave and allowed to have at it, heading down the coast. Border control was not what it is today.

Somehow a couple of us stumbled across this crew hitch hiking down US Highway 1. It was time for them to stop for the night. They were carrying a good supply of vodka for just such an emergency. We had a good supply of beer in the NCO barracks just because; plus the added advantage of a TV that would carry the game. GI ingenuity kicked in, so sailors from the enemy's navy partying and spending the night at one of the nation's protective radar sites seemed to be a good idea at the time. The next morning the First Sergeant found a bunch of his radar troops and Russian sailors sleeping off quite a night watching a hockey game on TV.

After a few tense phone calls, the powers at headquarters decided there was no underlying attempt at espionage or anything strategically untoward, just a bunch of GIs doing what GIs do left to their own devices. The sailors had gotten nowhere close to classified areas and were not the least bit interested in that stuff anyhow. The Soviet skipper suggested his sailors turn themselves around and head straight back to St. John's. The site commander suggested in rather strong terms that we never again pick up strange hitch hikers along US 1. For some time there were many little cleanup and other unappealing tasks that got done by a small group of "volunteers".

This was an interesting part of the world. In Machias, the train tracks crossed U.S. 1 by a food store that in the summer sold soft-serve ice cream from a window on the side. When log trains came through, the engineer would stop, run to the window, and pick up a cone. He would then move the train on; but stop when the caboose reached the ice cream. Locals understood and just waited patiently while the road was blocked. Drivers who got angry and honked their horns ineffectively were always out-of-towners.

Another thing that added spice to my life was the owner of the trailer park where I first lived. He was the Assistant Police Chief. The force was two people, so if there was ever a need for more, locals stepped in to help. He offered to take me on a ride-along one night. We drove into the heart of town and backed into a spot at the corner of U.S. 1 and Main Street. I asked Basil what our mission was, expecting a stake-out to catch some perps involved in some sort of misdeed. Instead I found that we had to keep an eye on a large box on the side of a building across the street. It seems that the freezer at the supermarket had been kind of dodgy of late. The bell on the alarm was broken, so we were to watch for a small red light in case the cooler quit so we could call the store manager.

To Sunny California

After two winters at Bucks Harbor, I received the assignment that changed my life and career completely; the 751st Air Defense Group, Mt. Laguna, California, east of San Diego.

By this time I had advanced up the NCO ranks. To take advantage of my broad experience, and because they had no one else to do the job, I was assigned to the Group Maintenance Office. In other words, I took care of all the paperwork.

My Change of Life

Through the years, I'd been picking up college credits and had gotten to within striking distance of a Bachelor's Degree. I have the old GI Bill to thank for that. While in the San Diego area I was able to take advantage of the universities there and get accepted for the Bootstrap Commissioning Program (BCP), one route for enlisted personnel to become a commissioned officer. BCP had relatively wide entrance requirements. If someone could accrue enough credits to where they could earn a bachelor's degree in 365 days from an accredited university they qualified. There were few restrictions on which degree to pursue, as long as it was a legitimate Bachelor's from a real school. I was in what would turn out to be the last Bootstrap class.

The other program, the Airman's Education and Commissioning Program (AECP), had more restrictive requirements and was designed to get officers with specific degrees, mostly in engineering or scientific fields from a school selected by the Air Force.

In my case, a Bachelor's Degree in psychology from The University of San Diego did the trick. I spent a full year where school was my only job. It was an excellent program and exhausting year. Timing for entering OTS (Officer Training School) was such that I had to go almost straight from final exams to Medina Base, Lackland AFB, Texas. An extra special result of this time was meeting my wonderful future wife Nadine.

How I Got There

Going from Master Sergeant to Second Lieutenant was interesting. Officer Training School was an adventure. While the majority of the officer candidates were focused on learning how to survive in the Air Force, my Technical Sergeant roommate and I spent most of our time messing with the instructors who had a grand time messing with us. Mario had family in the San Antonio area. Because OTS was explaining nothing new to us, we did not have to study the way the newbies did. So we concentrated on getting passes to visit off base on the weekends as much as possible.

After a few weeks, the officer trainees were, if their performance was satisfactory, allowed to go into town from after parade on Saturday until that night. For the other trainees it was a reward. For Mario and I, it was a goal. But there was a catch. Just to show us who was boss, the instructors began to gig us for all sorts of things. Demerits, real or imagined, for infractions of the rules or not being as squared away as a good OT should be could keep one on base.

But two old NCOs can be sly, devious, and cunning; they bear considerable watching. We discovered that we could earn merits by doing things for the school staff, countering the demerits posted against us. We found all sorts of things we could help the senior training staff with. We consistently did filing, wrote reports, etc., for the training office. We always managed to accumulate just enough merits to counter our demerits, so to town we went. Although there is no proof anyone would admit to; we had the feel that the command staff had figured things out and were somewhat complicit in the game.

One regular activity was helping Mario's grandfather evaluate his different home-made wines. He had a theory that the best way to do this was to cleanse the pallet with Tequila between wines. Before long Mario's wife and grandmother got suspicious that the wines were just an excuse for the three of us to spend the afternoon drinking Tequila.

One night we were invited to the Latin American Club. Officer Trainees had to wear their uniforms all the time, even off base, so this anglo stuck out like a sore thumb. At one point Mario and I had to stand up. There was some sort of announcement from the band stand which went far beyond the two Spanish words I could sometimes remember. I found out the announcer was explaining how Mario and I were in Air Force officers training. There was great cheering, applause, and drinks. One guy literally followed me into the Men's room to give me a beer. What a warm, wonderful night.

In the end, at OTS a good time was had by all and a couple of old new lieutenants were released on the world. It took a long time for me to get used to some things. Whenever I came in and someone called the room to "attention", I had a tendency to look around to see what officer entered. I also got the sense that when older NCOs saluted, seeing that I probably used to be one of them, they put a little extra snap and respect into it. I was officially a Mustang (worked my way up through the ranks to a commission.)

A New Career Begins

From Texas, it was back to Biloxi, Mississippi, and the Electronics Officer Basic Course. It was required; but in my case, major parts were a repeat of courses, experiences, and my jobs for the past decade. I zipped through the basic electronics portion that covered what I had been doing for a living by taking the final tests for each block of instruction, passing them with flying colors, and even finding a few mistakes in the questions along the way.

More School

Later in my career, I was sent back to Keesler for the Advanced Telecommunications Officer Course, the Air Force equivalent of a Masters Program. In fact, some students stayed on after graduation and attended a special year at the University of Southern Mississippi and graduated with a Master of Telecommunications Science.

One fascinating experience came during orientation. It was an international group of students and everyone had to come prepared to give a briefing on a particularly interesting event in their careers. In what was a real "holy crap" moment, the class was getting to know one another over drinks at a welcoming reception. Conversation turned to the briefings when an Israeli officer casually said he would probably talk about this "sort-of-special" mission he went on a while back. He showed us what he would be using to prepare – his field notes from the Entebbe Raid! This was the 1976 rescue of 103 hostages from a hijacked French airliner when it landed to refuel in Uganda.

My First Real Officer Job

On graduation, I received orders to the 4754th Radar Evaluation Squadron, (RADES) Hill AFB, Utah. The unit fielded teams made up of some of the Air Force's best radar technicians and engineering specialists to air defense radar sites around the world. In about two weeks, a team would align the radars to perfection and produce incredibly detailed coverage data for the site. An eval visit meant literally working sunup to sundown, using the sun as a hyper-accurate signal source to measure the antenna beam patterns. This was combined with data from several days of flying highly controlled test runs against the radars. Exact coverage would be plotted on a Radar Coverage Indicator (RCI) that became a permanent part of the site's technical records. This was important to understanding where coverage was best and where there were limits.

It was interesting to arrive at a site and start the initial briefing. After months of telex and letter communications setting up the evaluation and signed by a 2nd Lt. Seeing the looks on the site personnel's faces when they realized this was no regular "butter bar" heading up the evaluation. Once they realized I knew whereof I spoke, things went swimmingly. I also had the trust and confidence of my team, and I had confidence in them. They were the best in the world at what they did and the results were always good, although each evaluation was an adventure.

My career path and background could come in handy. There was one site on the northern border notorious for flunking evaluations. Before a team from the 4754th arrived, site technicians were supposed to have completely aligned the radars so all our specialists had to do was verify things and tweak the systems where necessary. An eval team would arrive, but the radars were so far out of alignment any attempt to do what they came for was futile.

The Division Commander was frustrated. He needed that site to be evaluated! He decided that I should take two of the best radar techs in RADES and "go find out what the hell was going on." The three of us dropped by Division HQ on the way up where he gave us carte blanch to do what we needed to sort the mess out. He also gave me a direct phone number that would be answered only by him to use as soon as I had a handle on what the problem was.

My team and I were not ten feet inside the gate when we could smell that the place was sick, real sick. Morale was terrible, there was no sense of pride, everything was being done in as sloppy a manner as possible. The site commander was massively unimpressive. Things would disappear unexpectedly. My team reported that the radar techs were qualified and capable, they just didn't care. One afternoon in the dining hall my gloves were stolen – while I was right there. When I brought it up to the supervisor, his response was "so??"

Time for the magic phone number. I called the Division CO and reported that the leadership level at the site was at minus the square root of bugger all and the whole problem could, as far as we were able to tell, traced directly back to one of the worst officers I'd even seen in the Air Force. The colonel thanked us and asked that we see what we could do about making possible a valid evaluation in the not-to-distant future.

By the next morning, the major had been relieved of his command and was packing his bags, the Exec was in command, and my guys were huddled with the site maintenance crews coaching them on aligning the radars. Six months later we sent a team in and the results were among the best ever reported from a site.

One of my more interesting trips with special meaning was taking a team to my old site of Bucks Harbor. By then the AN/FPS-24 had been removed and a newer AN/FPS-66 was installed in the old building, leaving lots of unused space. The new antenna looked completely lost sitting up where the old monster used to be. It was a pleasure meeting old friends both at the site and in town. An interesting thing about radar site veterans is how they developed a special bond and unique club. If you don't think so, see the way they swap stories and tall tales on the Air Force Radar Site Veterans Association blog. Site personnel also can have an impact on the local community.

In my previous assignment there I had become friends with the parish priest in Machias. Once he had to go to Bangor for Diocese meetings. I took a couple of days off and went with him. While there, we stopped at a basic used hardware/junk store and brought back a spotlight and timer, with no idea what we would do with them. We decided to plug the spotlight into the timer in the church attic and point it toward the round stained glass window. Setting the timer turned the light on when dark set in. While there for the evaluation I went down town and found locals were still gathering in front of the church at dark to "ooh" and "aww" when the stained glass window lit up.

After a few team trips into the field, I was given the job of Report Quality Control Officer and part time trainer for new officers. It gave me a feel for and understanding of sites around the world. I came face-to-face with the vagaries of radar and some of the wacky rules for handling classified material. In one case, Canada had requested that the squadron do an evaluation of a new Canadian radar at a Canadian site. The team did a good job, the report turned out excellent, and because of the NOFORN (No Foreign Nationals) classification restriction I could not send Canada the report. With a little finagling, we were able to get an "Except CANUKAUS" change to the rules, allowing us to work effectively with Canada, the United Kingdom, and Australia.

Florida Bound

As my time at Hill drew to a close, the Air Force decided they needed me as a Human Relations/Race Relations Instructor for the Social Actions Program at MacDill AFB in Tampa, Florida. My assignment was to conduct the Race and Human Relations Program classes that were required by the Air Force. On the way there I was cycled through the Technical Instructor Course, the Academic Instructor Course, and the DoD Race Relations Institute. It was an interesting time. Not only were classes on base, but I helicoptered to University of Florida, Gainesville, and visited our range facility in Avon Park.

The Avon Park facility was co-located with a minimum security prison. It was where I had some interesting experiences while there to conduct classes. The prisoners could roam about freely and did many things around the place, like cutting grass and other routine upkeep tasks. Everyone joked that they never had to worry about forgetting the combination to their safe because they could always find someone who could open it.

Only one prisoner ever broke broke out, according to the stories. He sneaked out with some laundry and was caught while trying to sneak <u>back in</u>. One prisoner refused every chance at parole. He said "I'd just get in trouble on the outside and end up in a regular prison. I have it good here." He worked in the Orderly Room, doing a great job and was very friendly and helpful.

Besides learning to deal with all sorts of people, I took advantage of a Golden Gate University night class MPA (Master of Public Administration) degree program. Instructors for the MPA program were practitioners in the field they were teaching and came in from their regular jobs to hold class. The Constitutional Law instructor had pleaded cases before the Supreme Court. Administrative Law was taught by the City Attorney of St. Petersburg. For his final exam we had to write suggested language for a city regulation relative to strip clubs. He admitted that he took the best language from our exams to write an actual city ordnance he had to come up with.

I received great encouragement from the Wing Commander. He pushed me to complete the MPA program before recruiters for the Air Force Academy visited the base looking for new teachers. But sometimes plans can be derailed by hidden details. Academy teachers must have at least a Master's degree. According to plan, and with the help of my boss and the commander, I graduated with my new Master's degree in time for the interview. It was great how everyone worked with me so I could complete my Masters.

The big day arrived and I went to my interview. When I announced that to the Colonel doing the interview that I had completed my MPA, a valuable lesson plopped down on my head. The colonel jumped up saying "you haven't graduated yet, have you?" When I proudly announced that yes I had, he let out a few swear words that roughly translated as "Houston, we have a problem". He let me know that I was SOL.

As it turns out, the job they had me slated for called for a Master's Degree in Psychology and the Academy would have sent me to the University of Chicago to get it. But because I already had one Masters, the rules would not let them send me for another one. Had I not gone to graduation and officially accepted my graduate degree, he could have worked something out. I was offered a position as an Academy Military Instructor; but I decided not to take it. Looking back, that probably was a good decision.

That sucked. But as the saying goes, "suck builds character". Apparently to make up for the disappointment the Air Force sent me to Squadron Officer School (SOS) at Maxwell AFB, Montgomery, Alabama. This was not the most exciting part of my career; but it was fun watching the Watergate gang that was incarcerated in the minimum security Federal Prison co-located with Maxwell and SOS, riding around the base in a truck full of mattresses. They seemed to be spending the days in the same truck with the same mattresses, no apparent work took place.

One interesting part of the assignment was getting to know a captain classmate who was a serious dressage rider and was working extremely hard to make the U.S. Olympic Team. On one strategic planning exercise, I came up with a recommendation that we nuke Canada. What really annoyed the instructors was that I was able to, according to the assignment rules, justify it. Gotta watch those Mustangs.

Head West My Man, Head West

While there, I received orders to Nellis AFB and the 4440th Tactical Fighter Training Group (Red Flag). This cold warrior and air defense weenie at first had no idea what Red Flag was all about. Then someone explained that it involved shooting down airplanes all the time.

Cool! The rest is history.

Adventures Along the Way.

During a career, many in the military find themselves caught up in situations well beyond their duty assignment. Helping the local community in time of need comes naturally. All the traveling meant that they often go to a disaster, not the other way around. It makes for some intense memories. I found myself involved in such situations more than once.

In 1969, while at Keesler AFB, Hurricane Camille struck and devastated the Gulf Coast from Pass Christian to Pascagoula, Mississippi. At the time, I was living in a mobile home just outside the Pass Road Gate, not a good place to be when a severe hurricane hits. So I took the family to stay at a friend's house in Ocean Springs, further away from the Gulf front and in a far more substantial building than I had.

To make a long story not as long, Camille turned out to be far more severe and destructive than predicted. It wiped out the bridge across Back Bay, our route back to Biloxi, and took out all power and communications. As a result, like many other Keesler personnel, the base had no idea where we were and we had no way to let the base know we were OK. So for a couple of days all of us in that area were listed as "missing" on the Air Force rolls. Eventually, someone managed to collect a list of names and get word back to base about us. Everything was shut down and it was nearly three days before the flood waters subsided enough that we could make our way back to base, going a very long way around the water.

The Gulf Coast was devastated. I figured my mobile home was a goner. But when I got back to the trailer park, there it was, just like I left it. The only damage was to a metal storage shed that was filled with a lot of stuff all stacked up. It was amazing. After all that wind, the stack of stuff stood just like I left it, except the shed itself was gone, probably blown half way to Hattiesburg for all I knew. Except for a bicycle that was next to one wall having fallen over, not a thing was damaged. That included Christmas ornaments.

The Gulf Coast was ripped apart and a storm surge wiped out everything from the water to the railroad tracks blocks inland. Even those of us who knew the area intimately could not recognize anything along Highway 90 all the way to Pass Christian. It took Hurricane Katrina nearly four decades later to inflict more destruction on the area.

Although Keesler was not an active airfield any more, it still had a usable runway and flight line; the former last used by a resident Coast Guard Detachment, the latter used for student parades. Regular school operations were out of the question after the storm, and would be for some time. The area needed help, and we had the only place planes could land. So Keesler AFB, the Electronics Training Center, became an active Air Force flight center for the duration of the emergency.

The military always steps up in time of need, often being the only organization with the equipment, personnel, and ability to get things to where they are needed. In this case, the personnel (the school student body) were already on scene. They were put to work on the most pressing need, helping clear debris and searching for survivors (or victims). There was a dire need for heavy equipment; trucks, bulldozers, and tools. That would come in from bases around the country via our runway.

This turned the ramp area into quite a scene. For the first couple of days, aircraft, mostly C-130s, came and went round the clock bringing stuff in as fast as they could. Those of us with truck driver licenses were assigned to help unload the vehicles as they arrived. A C-130 would pull into a wide part of the parking apron and open its rear ramp. As the load-master released the tie-downs, one of us would hop into a truck and drive it off the plane to a marshaling area on the grass. The planes did not shut down their engines to save turn-around time. As soon as the load was off, the pilot would taxi out and take off to go get another load. Just about every base in the southeast was providing stuff. The pilots were operating so fast that one actually began his taxi before I had the truck all the way off the ramp, literally pulling out from under me.

The military family has a big heart. Every truck that arrived carried boxes of stuff donated by the folks back at the base of origin. Clothes, shoes, household items, all direct from those who had to those who suddenly had nothing. One thing hit me. Sticking out of a cardboard box on the seat of one truck was a well worn, well loved stuffed bear. A child somewhere sent a beloved friend to comfort a needy child he or she did not know on the storm-ravaged Coast. At the end of one night's operation we ended up with a 40-ft trailer stacked to teetering with boxes from families at bases all over.

Days into the operation, a commercial 727 surprised us by landing. In the '60s, the runway served T-28s when Vietnamese pilots trained there for a while. Somehow the pilot got stopped just short of Back Bay. It blew our minds when we started unloading the plane and found it filled with case after case of Schlitz beer! We popped one open and found water. Drinking water was a pressing need, so someone at the brewery creatively changed the production line, filling the cans with drinking water – later surprising the hell out of a bunch of sweaty GIs unloading an airplane.

We were beginning to think Keesler Air Force Base was going to have a new static display in honor of Camille activities. But a week or so later, the Gulfport airport re-opened and the pilot took on just enough fuel to get the 727 the roughly 20 miles. It was quite a show, backing the airliner until the tail touched the boundary fence, spooling the engines to 110 percent, releasing the brakes, and going for it. We swore the last few feet of the takeoff roll were skittering across the surface of Back Bay.

Keesler was awarded a Presidential Unit Citation for efforts in the days after Camille.

I learned one thing about responding to disaster victims. Good hearted people want to give, sometimes literally the shirt off their backs. But this sort of generosity can cause unintended consequences. Months after the storm, there was a warehouse filled to the rafters with donated clothes that had not yet been sorted. Even opening up and letting needy people help themselves we could not give everything away. By then there was much less need.

When charities ask for money, instead of donating goods to help a disaster area, it can be easier to have cash and buy just what is needed and have it delivered direct to where it is needed instead of generating piles of stuff that become unmanageable. Cash can be focused on specific needs, a more efficient way of using donated resources.

After the trucks were delivered, we set about hauling stuff. One morning about 3am, I made a run to Gulfport with a loaded 2-ton 4X4; getting there in a dense, dripping fog with my rear end really dragging. It seemed unworldly, the opening for a horror movie. Out of the mist came a voice, "hey Sarge, how about something to eat?" There, in a Salvation Army trailer was this guy with hot coffee and ham sandwiches. Other relief agencies were still setting up air conditioned trailers before they would start helping. But here was the Salvation Army out in the debris feeding workers who really needed it. Since then, the Salvation Army is always at the top of my donation list.

Once, while driving alone cross country between New Orleans and the West Coast, I pulled up for the night just outside of Wichita Falls, Texas. The weather was looking dodgy, at best. The couple that owned the motel said I could have a room that was on a standing reservation for a sales rep they were sure would not show up because of the weather. It was on the contingency that if a storm hit that bathroom was their tornado shelter.

Sure enough, in the middle of the night the sirens went off. The couple, their dog, and I spent most of the rest of the night sheltered in the small but strengthened bathroom. When things calmed down, we went outside to find out that a tornado had destroyed a major part of Wichita Falls. There were still severe storms around; but the worst was up the road.

When something bad happens, the military, like first responders, run toward, not away from the trouble, thinking first "what can I do to help?" So when I found out that the National Guard had been activated and the armory was a few blocks away, I headed there. Identifying myself as on leave, on the road, and ready to do what I could, I was put to work helping them install new batteries in their radios and checking them out. When the unit headed out it was a satisfying feeling seeing the trucks roar out of the door, knowing help was on the way for the victims of a terrible storm. And I helped that happen.

I headed back to the motel. By then, survivors were beginning to show up from the stricken areas. Everyone at the motel did what they could, which included going through suitcases and digging out clothes to replace the torn and bloody things the victims had on. That was one hell of an overnight stop.

The Mount Laguna radar site sits on the eastern edge of the mountains inland from San Diego. It had a commanding view of the Salton Sea and the Navy's only mid-desert submarine base. To the west was a forest running to El Cajon, California, and south into Mexico.

One day, one of the fierce, hot winds that sometimes blew in from the desert and took down a power line a couple miles from the site. We turned out, but the fire had taken off through the forest before site personnel could do anything about it.

The Forest Service began a major fire fighting operation, calling on Mt. Laguna personnel for help. We were supporting the fire fighters communications net, leading to some interesting experiences. I was driving a fire captain to the command post. The access road was cut along a mountain ridge. We came around a corner just as flames crested the ridge line above our heads. Discretion being the better part of valor, we decided to turn around and get the hell out of there.

(Source: U.S. Forestry Service)

By the time we turned the truck around, the flames had rushed past us, jumped the road, and was racing toward the little mountain town of Pine Valley. Fire trucks had been deployed to the edge of town, ready to deploy as needed to save what they could of the buildings. But the wall of flames roared through Pine Valley with such speed that the only damage done was the corner of a porch burned. The fire continued for days, becoming the largest burn to date in that part of the country, reaching all the way into Mexico.

Some events become memories that never fade. One was putting the coffin of an airman who had been killed in an auto accident on the train home. It was the dead of the night at the Biloxi station and a chilled dense fog wrapped everything and penetrated to the bone. I will never forget our group standing by the tracks and holding a salute as his last train home pulled away. Once the red lights faded into the fog, we headed to our NCOIC's house on base for a badly-needed drink.

Keesler sometimes drew on students in the Telecommunications Staff Officer Course for occasional outside duties. Mine turned out to be leading a burial detail for a WWII veteran in a small country town nearby. The tiny church in a rural part of the county probably dated back to the sharecropper era. The community surely did.

We got there early, having allowed extra time for our bus driver to navigate the back roads to get there. The old vet was like so many from that war, a hero whose exploits were written nowhere but in the hearts of his friends and family in this little town.

Everyone at the church was apologetic at taking us away from whatever it was that we would have been doing back at the base. We, on the other hand, were honored at being able to do this one thing to honor an old soldier who had probably received little honor during his long life. We were also nervous that we would get everything right in the ceremony.

A wonderful, heartwarming part of the day was the result of the church ladies' insistence that "their Air Force boys" be treated right. The basement of the church was filled with the delicious smells of their cooking. By the time of the ceremony we were filled with some real good Mississippi country cooking lovingly prepared for us.

After a back-country, black church funeral, we could feel the cotton field history in the song and prayers. After taking the flag-covered coffin from the church we headed for the cemetery. A big worry was getting the old blue bus up the narrow, rutted road and between the slash pines to the family cemetery in a small clearing.

We got through to the grave side ceremony successfully. The flag got folded properly, the American Legion octogenarian honor guard got their salute shots off mostly all together, and I was able to present the widow with the flag "on behalf of a grateful nation." She was touched, and so was I.

The ride back to Keesler was quiet. Saying goodbye to a true old soldier touched us all. We did not know him, but would never forget him.

(Photo: lehmanreen.com)

4. Life In Red Flag

It was the best of times, it was the worst of times. My tour at Red Flag was by far the best job of my career. It was the most challenging, the most interesting, most satisfying job imaginable. It was a lot of hard work and crazy hours; but I never regretted a minute of it.

Some things were not fun. It was heart breaking to stand in the middle of a smoking fighter jet crash holding the deceased pilot's wallet to confirm his identity. It was a wrench seeing something go terribly awry and the plane dive nose-first into the desert. Up-range there was always the danger that a confused pilot would bomb a simulator site, mistaking it for a target. One morning threat operators discovered a 500 lb concrete bomb embedded right at the step of a PortaJohn path,

But it was always gratifying to see crews' improved performance by the end of the Red Flag, to receive the appreciation from helping a unit discover a way of avoiding death from a surface-to-air missile, and to hear years later that many of my efforts helped create major improvements to Red Flag. To this day I cannot read about how the debriefing system is so valuable without a flash of pride.

Here I Am

I got to Nellis with a feel for life in the fighter community because of my assignment at MacDill AFB in Florida. It was a vastly different world from my career in Air Defense Command. I'd grown used to jets, their noise and smell, the ebb and flow of a tactical base. I'd gotten used to life among those who fly and fight for a living.

But Nellis, from day one, was different. Coupled with the fact that the base was in Las Vegas, a place like the back side of a strange planet, I knew I was in for one wild ride. What I did not know was a ride on what?

In the beginning, I stayed at the Visiting Officers' Quarters while looking for housing and getting settled in at the 4440th. With time on my hands in the evening I tried to learn everything I could about where I was and would be for a while.

One of the first things I came across was an article in the *Las Vegas Review Journal* about the chef at one of big casinos who was given an award and big bonus for managing his restaurant. He lost $70,000 that year; exactly on target! What kind of place was this I wondered? Then it began to dawn on me; casinos advertised "The Loosest Slots in Town! 97% Payback!" Making people feel happy with cheap food and free drinks is what it was all about. Visitors probably wouldn't notice is what the signs were really saying was "when you play here you'll lose 3% of your money."

Later, a friend who was the Purchasing Agent for The Frontier Casino (where I got engaged, but which has since been imploded and torn down to be replaced by a gigantic casino hotel complex) put this in perspective. I remarked about a newspaper article about a visitor hitting a million dollar jackpot on one of their slot machines and how the casino took such a big hit. She then explained the secret of Las Vegas – "In the time it took to make a big show of presenting a giant winner's check and taking a picture for the newspapers, we made that much just in the nickel slots."

What a town!

The Big Doors

On the way to work the first time, when I stepped out of my car a new smell hit me. I was familiar with the aroma of burned jet fuel at airports and MacDill. But this was different. It was the smell of massed fighter jets, the smell of combat; but a special kind of combat. The air was charged with purpose and my skin tingled with excitement from the feeling that there was no other place like this on the planet.

I was now a Red Flagger and the top 'bad guy'. One thing I worried about was how a non-rated Captain would fare in a world full of super-charged jet jocks. It was not long before I got my answer; "just fine." Rank determined my pay check. What I did and how I did it would determine whether or not I was accepted.

I was the head of Threat Analysis. Being a captain was secondary to the job. I quickly had it explained by the Red Flag Commander that if I shot someone down, they were shot down, no matter their rank, lieutenant or colonel (I never got to shoot at any generals). It was not long before I found out that from lieutenants to colonels were OK with that. That's how Red Flag works.

This would be life in the tactical world; but on steroids. Here I was, a 'Mustang' Captain with ten-plus years enlisted time, making it to Master Sergeant before getting commissioned and usually working about two pay grades above my pay check. I was suddenly thrust into a flyer's world. But not just any flyer's world. Red Flag was the top challenge for a fighter pilot.

The environment can be unforgiving and tension-filled and there were no bragging rights for second place. It was not long, however, that I began to find my place and was accepted by the pilots and crews for what I and the threat range could do for them – help them learn to survive.

There was thrill and trepidation passing through those doors and walking the hallowed Red Flag halls for the first time. These were the same doors the crews walked through going to and from the flight line and their chariots of fire.

"If Nellis is the home of the fighter pilot, then the fighter pilots' playroom had to be Building 201, Headquarters of the 4440th Tactical Fighter Training Group." (*Red Flag, Air Combat for the '80s,* p.53) It would be the Red Flag players' base of operations while at the exercise. Maintenance teams worked out of hangers across the street and their aircraft parked in a dedicated area of the ramp.

Out of the Big Doors (Photo: U.S. Air Force)

It was interesting to watch aircrews fully suited up with their parachute and other flight gear swagger through the front doors. Suddenly their mood changed. As the standard exercise briefing pointed out, "No matter how much you planned, once you pass through those doors, you are not going to get any smarter crossing that parking lot." This was serious business they were about to do. They were going to war.

As Red Flag Staff, we organized the range capability and made it effective as a training medium. Based on first impressions, I felt like I was facing the challenge of running World War III/IV/V all at once. While this is something of an exaggeration, there were days it would seem true.

Knowing that you are helping run the place where air crews practice combat in a live, realistic environment was exciting; and more than a little scary at first. Pilots had the chance to try new ideas and work with other forces in a safer environment than actual combat. We had to make sure there was nothing that made things more dangerous than necessary and that the system let them make the best use of the range.

Home Sweet Home

Red Flag staff enjoyed a position of respect around Nellis. 57th Fighter Wing, Fighter Weapons School, the Aggressor Squadrons, the Thunderbirds, and Red Flag made Nellis a special place.

(Photo: U.S. Air Force)

Even though it was a fighter pilot's world, the Red Flag ground pounders enjoyed a status non-flyers elsewhere sometimes did not. "Other Agencies" operating up range and at special places around the complex, added to the mystery and 'specialness' of the place. USSR-ish patches and pins made it kind of fun being a bad guy. Interestingly, while Red Flag flying staff could be the focus of participants' ire, the threat analysts and threat operators got respect after missions. Instead of complaining about mission tasking or evaluations with the flyers and intel people, participants were more interested in picking our brains for how they could avoid getting shot down.

Uniforms could be somewhat creative; adding to the unique feeling of the exercise, like costumes in a play. Up-range crews wore jungle fatigues rather than the standard greens. Baseball caps reflected their range 'bad-guy' role. This is when I picked up my tactical call sign "Frenchy".

At one point, someone made arrangements to buy some real USSR Red Star hat pins for the range operators. They loved and cherished them. Mine was on order when the powers-that-be squelched the order from the (then) Soviet Union. So I never got mine.

I ended up wearing jungle fatigues most of the time. We came up with a special patch with the Tonopah EW Range and 554th Range Group in Cyrillic lettering.

Red Flag/EW Range Patches *(Photos: Kernan Chaisson)*

One day, the Red Flag Operations Officer sent me to get a flight suit. This was before my first Maple Flag exercise in Canada. Being non-rated, there was some doubt about whether or not this was authorized. But nobody said anything against it, probably because it was Red Flag. I used my 'goat bag' from time to time the rest of my days in the 4440th. I adapted to a standard that allowed me to wear either my blue flight cap or a Red Flag baseball cap. At one point we tried the old-style round box fatigue hat from the '60s. That was not terribly successful, although it would have been a great place for the Red Star I never got. I never ran into any questions about non-standard uniform.

73

Our Digs

During my time at Red Flag we were crammed into an old cinder block building that did not look like home to a world-class operation. Headquarters, personnel, survival equipment, intelligence, threat analysis, aircrew planning rooms, the briefing auditorium, a flight surgeon's office, and the all-important Red Flag Snack Bar (serving some of the greasiest, most delicious hamburgers on the planet). Events there were often likened to the bar scene in a *Star Wars* movie.

What didn't fit in the main building, and that was a lot, went out back in a jumble of sparsely furnished mobile homes. This included the Desert Survival School Detachment, easily spotted because of the snakes, lizards, and other creatures in cages at the front door and inside.

One of the Survival Detachment Residents and Adjunct Instructors (Photo: U.S. Air Force)

The Detachment's exhibition of snakes and other desert creatures drew a lot of attention, fascinating most participants who would visit their trailer out back frequently. Charlie the Burrowing Owl for some reason once took a great dislike to a particular bald-headed major; swooping while in full flight and nailing him with his talons on the top of his head. Otherwise, he was a pretty peaceful guy, just hanging out watching visitors come and go..

Having a Flight Surgeon clinic in the building came in handy. The Brits came in with their own that sometimes helped us by doing minor treatments so we did not have to get involved with long waits at the base clinic for treatment or prescriptions – including slicing off an annoying mole. Red Flag staff were also given dental appointments at the flight surgeon dental clinic on base, assuming that everyone was rated and needed the extra special way work was done to insure fillings did not pop out while at altitude. To this day, some of the fillings I got while there are still in perfect condition, to the amazement of my regular dentist.

Foreign participants made their headquarters out back also because they often needed more space than a room in the main building could provide. It took larger contingents to support their operations. It was interesting to see how some of the nations personalized their little home across the pond for the time they were there.

Participating crews found their accommodations in the 'RF Mobile Home Court'. SAC (Strategic Air Command) Liaison had a home in another trailer. Since there were typically only a few missions a week coordination was simple because the B-52s used the same basic ingress and egress routes and a set group of targets;. The only thing different were the bomber crews. As a result, the SAC folks had time on their hands. One case was a Captain who spent time hatching ideas for a book that would take advantage of his B-52 experience. Dale Brown did some early writing while at Red Flag as the SACLO (SAC Liaison Officer) coordinating B-52 missions over the range. One day he told me he was working on a book where a B-52 was the main character. We know how that turned out.

His experience created a unique character for a series of popular techno-thriller novels introducing a unique character, the B-52 known as 'Old Dog.' This future best-selling author also took advantage of his efforts to keep inexperienced crews from running afoul of flight restrictions involving the place "that did not exist" but is the target of all manner of crazy UFO and alien theories. There were some locations in his novels that seem a lot like *Area 51*.

The Neighbors

Next door was Range Group Headquarters and the Operations Center blockhouse. At the back corner was the microwave tower that supported the big dish antennas of the range microwave communications system. This would play a major role in future, crucial enhancements to Red Flag and the range. The bright red-orange ASR-7 Airport Surveillance Radar rotated on a platform behind the building. It would, through some sleight of hand with authorization documents, be upgraded to an ASR-8. The ASR-9 slated for Nellis went to Tolicha Peak where the better radar would support range missions more effectively.

(Photo: U.S. Air Force)

Further down Flight Line Road was Headquarters for the 57th Fighter Weapons Wing Headquarters as well as Fighter Weapons School. The two *Aggressor* squadrons were easy to spot by the hammer and sickle on the front of their buildings and everywhere else, including in the latrines. The pilots were particularly proud of the way a latrine looked with every seat raised showing their special contribution to the overall décor.

The far end of the street is where another highlight of Nellis made its home, the Thunderbird Aerial Demonstration Squadron.

(Photo: U.S. Air Force)

This place could be strange, unusual. Before the new hospital was built, there was a clinic near the flight line. The medics were proud of a new setup that connected to a remote computer at a distant hospital which analyzed EKG results and sent back a report. Afterburner takeoffs raised holy hell with the EKG machines, though. More than one person was declared 'dead', or at least 'extremely critical' by the computer during a physical because a pair of F-15s did a max-power takeoff.

Nellis never lacked for something noteworthy going on. It was always interesting to see 'Janet' flights on the other side of the flight line. There were also special flights loading and unloading at the end of the runway, near the engine run-up cells. No one asked any questions.

The Out-of-Towners

Maintenance crews relished Red Flag as a chance to show what they could really do in a deployed, field situation. They worked exceptionally hard to be sure their airplanes were ready for all missions without problems. They were very creative when things broke. Lights burned and activity continued all night in the hangers.

Red Flag participants had a way of creating memorable situations. One came during a Marine Squadron deployment. There was a bar outside the Main Gate, a nice little neighborhood place that base personnel frequented. The Marine maintenance crews made it their 'field headquarters' for the duration of their stay during one Red Flag. One night spirits ran high. Nothing mean or combative; but at some point during the night it appears that the group got into "I can top that" mode and, to make a long story short, one of the urinals got ripped from the wall. Don't ask.

One feature of Red Flag was the tendency to let units handle situations on their own. In this case, the RF Commander gathered the group together and explained that the damage should just go away without any Red Flag involvement. Sure enough, within a couple of days the Men's room was as good as new. Actually, that was a nice improvement. In appreciation and to memorialize the event while honoring The Corps, the owner installed a beautiful brass plaque dedicating "The Marine Corps Memorial Pisser." It probably stayed there until the building was torn down a decade later to make room for a new base hospital. I hope someone thought to save the plaque and send it to the Marine unit.

Red Flag had an active intelligence operation to exercise a participating squadron's built-in capability. At the start, they would establish scenarios and evaluate the participants intel operation. It would be interesting how, at first, the squadron tended to report back before they had a feel for the situation. Like so many operations, the first reports were too hasty and information too light. With staff guidance over the two weeks, unit intel shops became more effective, much better supporters of the unit mission. In combat this can be a life-saver.

The issue of participant progress is another interesting thing. In the first few days, the Red force and threats usually had their way with the participants. Bragging rights to the bad guys. As the missions progressed, the visiting teams began to get the idea and got better. By the end of the two weeks, things had changed drastically. The Blue Team began to work together, stopped making rookie mistakes, and began to have their way, sort of, with the home team. This always led to lots of noise in the snack bar as they laid claim to bragging rights about how they were now almost as good as the Red Force and could kick butt any time they were called on to do so.

We could see that they were not the same bunch that landed two weeks before and most of their bragging was justified. They were ready to do their job now. Some bought beer to help us feel better about getting beat upon by them. The participants' being able to take Red Force to the cleaners, or at least get good enough that we no longer mopped up the floor with them, is exactly why we did what we did; and it took Red Flag a lot to hard work to get to where this was always the result. While the pilots were congratulating themselves at the Snack Bar, behind the scenes it was high-fives among the Red Flag staff. We did it again!

A Day in the Life

The Threat Analysis day started early, real early. In fact, it started the previous evening up range. Once the afternoon Red Flag mission finished, usually mid-afternoon, the threat operators had the task of correlating, processing, recording, and getting the day's results ready for Threat Analysis at Nellis. This involved gathering the tapes and operator sheets from the treat sites. Supervisors and operators viewed the video tapes and operator logs, making sure everything correlated and would be usable for scoring and debriefing the pilots. Any special circumstances or attention-worthy items were noted.

This was where the results of EW on an attack package were first evaluated and the operators were the only place this could come from. The up-range operators were the first to see the attacks and generate mission results. They were the most valuable link in the EW range chain. In those days, the ability to collect, record, and evaluate electronic combat's impact automatically did not yet exist.

Everything came from the operators' good old class-A1A eyeball as well as knowledge and experience. They were good! This was one of the most valuable parts of Red Flag training, something pilots could not get anywhere else in the world.

Next came the job of preparing and packing the tapes and logs for transport to Red Flag HQ, a task made tedious and complicated because most of the information was classified. The trip from the range to Nellis could, given the normal lack of traffic overnight, could be made in three or four hours. The drivers did a great job and in the wee hours the tapes and logs made it to the Threat Analysis safes ready for us to retrieve, also in the wee hours.

During a Red Flag, my alarm rooster typically rousted my rear from the bed around 3:30 am. While JP-5 was fueling the airplanes that dueled in the skies above the range, caffeine is what kept the ground threat operation going. As Chief of Threat Analysis, my first job was to dig the stuff out of the safes and make sure we had everything we needed for the day. For most of the time there was a second officer helping with analysis of the daily tapes. Although this would be the basis for the afternoon debrief, we had to be ready to talk to crews one-on-one about the previous day's missions as soon as we could.

Each analyst position consisted of a TV monitor (not a nice flat screen, but a huge square box that weighed a ton and could be very moody, especially annoying when we were in no mood to put up with a moody television). There were two heavy duty commercial videotape recorders that took ¾ inch cassettes. One was used to play the tapes and the other to record one for that evening's debrief (and those damn things could be even more moody that the TVs). We usually operated with the covers off because we had to frequently wipe off the play heads with a special fluid, which was probably bad for us. The player had to stay clean so the video did not lose lock and become scrambled as we tried to watch it or cross-record it. The tapes could pick up a lot of fine dust up range, dust that messed up the video. Whoever invented the thumb drive should be made a saint.

Dubbing representative video clips for the mass debriefing entailed watching the video and finding a piece that would be especially good to show; then with delicate flying fingers cue *play* on one and *record* on the other tape machine to dub the clip. The machines had a built-in delay as the play and record heads spun up to operational speed and the transfer would start. Naturally, just to add spice to our lives, each machine was different.

It could be frustrating, so say the least, to produce a debriefing tape of the quality we wanted for the debrief. The process also called for an ability to see into the future. Is this the best of what there is, or will something better come along? If something better did pop up, in those days you could not go back and just insert it. New video could only overwrite what was already there. Everything from then on had to be re-done. That could be a pain, and there was no assurance the new clips would record as well as they did the first time.

Folks in the outer office knew when a serious dubbing operation was going on by the muffled obscenities coming through the closed door of the video room. Today, there are all sorts of video systems that can be used in many ways and for producing dubs of parts of an original. For a few hundred bucks at Radio Shack or other electronics stores one can pick up a machine that will easily and quickly do what we struggled with in the wee hours back then.

The other analysts and myself routinely asked Santa Claus for the ability to analyze and debrief missions right after they landed, while the feeling and memories were fresh and time had not dulled the immediacy of the experience. That is when critique does the most good and what pilots learned could be applied to that or the next day's missions. I did not realize that in a few years I would be installing the communications system that would make it possible for these wishes to come true. The resulting system would become what pilots would call one of the most valuable parts of Red Flag.

My goal was to get the threat de-briefs ready as soon as possible. The aircrews goal was to start aggravating us for results a half hour earlier than that. I created a sign to hang on the video room's door.

"No, THE TAPES ARE NOT READY. Do a quick 180 and go away. Watch this space for another sign that says THE RESULTS ARE READY." Most of the time this almost worked. Actually, how eager the crews were to see the threat results and discuss what happened was a tribute to what they felt we and the crews could do for them.

The Value of Getting Shot At

While the flying and fighting was fun, crews really appreciated the ground threat part of Red Flag training. Operating against live ground threats was an opportunity they would not get again until, or if, they came back to Red Flag. It was rewarding to work with fighter pilots that were considered by many as the top of the Air Force, and feel their interest in and appreciation of what we did. At that time, it was not flyer versus the ground guy, it was everyone doing their job and bringing the results together so tomorrow's mission or, should it ever happen, combat results would be what should be expected from the United States Air Force.

There was always a line of pilots and electronic warfare officers to work with. I also had to deal with a steady stream of questions from the data entry office about what entries should be, queries from planners about ways to set up the range operation to better simulate particular missions, plus all sorts of other questions that the questioners could have figured out for themselves had they bothered to think about it for a minute. When one is seen as the only one doing a particular job, everyone else seems to assume that you are the only person who can answer their earth-shaking question, no matter what else you have to do. Oh well.

During a mission 'go' my job was to monitor and coordinate the up-range operation in accordance with the mission plan of the day. This included making sure that the threats attackers saw were both technically and tactically accurate for the mission assigned. The range display could display lines from select threat sites when they electronically fired. This timing and direction, coupled with the electronic ID of an aircraft, gave me more information to use in a

"God's Eye View" picture of missions as they were happening. This was useful in being able to give a more complete evaluation of what was included on the threat logs and videotape.

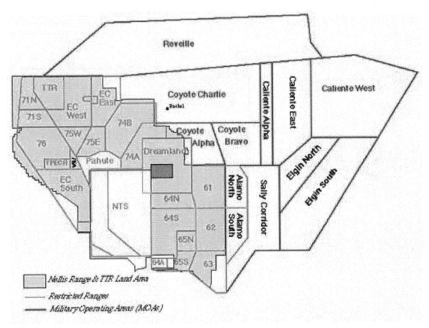

The Range. Most of the threats were in the upper left quadrant.

Bang, Bang; You're Hit

How the ground operation went about protecting targets could vary. How an airfield was shielded was different from how the shooters threw up a shield for an industrial complex. Convoys looked different, from an electronic order of battle standpoint. Because most threats were fixed in place, we had to tailor the way we operated to boost the training value of the encounter.

While the ground threat operators were given a general mission plan for each set of sorties, there was the need for real-time coordination of their efforts. Being linked with them was also a way of making real-time adjustments based on last-minute changes to flight orders, clever mission changes based on what was learned from the Threat Analysis debriefings, or from some too-clever-by-half idea detected during work with crews during the de-briefs. While it was one thing to tell a crew that their idea was bone-headed, it was better to help them learn for themselves in the ground threat video and engagement sheets. At the time, Red Flag was the only place this was or could be done.

I would operate out of the Range Group Control Center using the Weapons Controllers' Range Control display, calling engagement instructions to the threats from a small dais on the edge of the control room. There was a limited early warning net up range, so I would feed instructions to the operators and coordinate their tactics as though I were part of a real air defense net. Tactics could be synchronized with the scenarios set up by the Red Flag intel shop, and we always tried to give crews a different look from mission to mission.

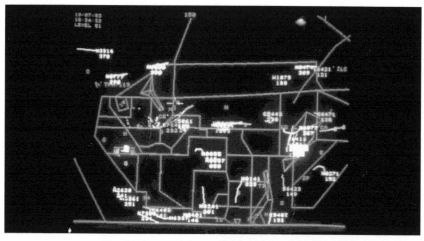

The Big Board   ***(Photo: U.S. Air Force)***

It was also an opportunity for some special work, such as helping units evaluate specific tactics or, as happened with a Squadron of special mission C-130s from the RAF, help them develop a way to

84

counter particular vulnerabilities. They needed to find a way to avoid being killed by a particular threat using their available protective equipment, which was none. During the first few one-on-one threat de-briefings the crews and I caught some comments on the video from the threat operator, leading to a first-class head-bang among the pilots, their flight engineers, and myself.

We came up with what everyone thought was a dumb idea that could never work. So naturally we went to the planners and got the RAF portion of the missions isolated from the busiest part of the attack flow so we could try a few things. Based on the fundamental concept of scientific investigation, "poke it with a stick and see what happens", we used my conference net with some range operators and a radio link to the flight engineer on the aircraft and discovered a way to generate some rudimentary protection from a worrisome and deadly threat.

Red Flag was continually the source of some special, personal rewards. In this case, I was the only Yank invited to the Brits-only post-mission debrief back in their trailer featuring St. James Best Bitter, a favorite. I also have a small plaque on my wall with the squadron coat-of-arms on it. From what I have been given to understand, there are very few of them living anywhere outside the U.K.

Range Control was always a happening place. Range time is like gold. When Red Flag was not using it, everyone else wanted to. It was not unusual for the front of the first Red Flag mission to enter one side of the range just as a Fighter Weapons School mission exited the other side. It was always fun listening in on the Controllers while setting up for and sorting out Red Flag range time.

There was one chilling experience when the dreaded "knock-it-off, knock-it-off, knock-it-off" blared over the emergency frequency speaker followed by "no-chutes, I repeat, no-chutes." You don't have to be a flyer to know that the day had just turned to crap. To make things worse, a well-liked, very skilled Weapons Controller was in the back seat of one of the F-15s, on the mission to get familiar with dog fighting from the other side of the scope.

The planes were involved in a mid-air collision where one slammed bottom-first into the cockpit of the other, killing the pilot and controller instantly. Listening in on the fatal event was terrible; but when a co-worker was involved it could be excruciating – especially when we had to begin the data collection, rescue/recovery, and investigation operation. In Range Control, in spite of having just lost a friend and colleague, everyone did a quick, joint inhale, let out their breath, and immediately became a well-oiled machine taking care of business. Grieving would have to wait.

New Gear

The Air Force was always cooperative when it came to upgrading and enhancing our jobs. One day, I received a new Plantronics headset in place of having to use hand-held phones or bulky earphones and microphone during the missions. They were great and the newest headsets on the market at the time. Small, light, comfortable. I especially liked the tiny tube microphone that wrapped around in front of my mouth. I saw one being used by and astronaut once and thought it was neat. (That's the way we talked in those days.)

It took some doing to get the things adjusted comfortably because of my glasses, so I would just leave my headset on between the morning and afternoon missions. It made me look like a jerk; but I sort of used it to show that I was a special operator and support an ego surge that said "I have one of these and you don't." It also advertised that "I am the one who can shoot you down." Red Flag did not always bring out the best in people.

Battles with a strange system

It is interesting and sometimes puzzling how the system operates and had a unique way of taking a good idea and turning it upside down. The threat analysis office needed new TVs for processing the simulator video. We found Sony sets that met all the requirements and at a good price (around $250). But after they were delivered, someone in the supply chain decided that Sony TVs violated a "Buy America" provision of the procurement rules. They replaced them with what appeared to be identical sets, except for a simulated brass plaque on the back with supply numbers and "Made in America" printed on it (at $500+).

Because the sets looked so familiar, on a suspicious whim I decided to pry one of the labels off. Guess what. Embossed into the plastic case under the plaque was the Sony label of the original sets we bought. These were the exact TVs we originally bought but were forced to replace. I guess the procurement rules could be satisfied by gluing fake metal labels to the sets we wanted, and adding $250 to the price. The "Made in America" label was probably the only thing made in America - maybe.

Morning mission over, time for a break, right?

Threat operations for the morning operation were usually over by late morning. The second 'go' would not get to the range until mid-afternoon. That gave me a few hours to work up as much of the previous day's scoring and select video for the mass debrief that afternoon. By then, the other analyst would have helped me pre-select material for the daily dog and pony show, as we sometimes called it. While there were many, many wonderful lessons buried in the results, I had only a few minutes on the stage. So selection was important.

Preparations for the afternoon were needed. Intel shop last minute mission changes did nothing to simplify things. Unlike the air-to-air fight, the ground threats sometimes had no way changing to keep up.

Invariably, a pilot returning from the morning seemed to think that overnight some miracle made it possible for me to give him the results of the just completed mission. "Can you at least let me know if I got shot down?," was the hope-against-hope plea. This would have been valuable; but unfortunately impossible. The day would come, however.

Wanna sit in the desert instead of flying?

One of the more interesting tasks I had was picking a "volunteer" for Search-and-Rescue Exercises (SAREX). I usually did this from the threat kill sheets. Visiting squadrons were given consideration and could recommend crews that needed their survival training updated. But there were times when other factors came into play.

Just because you are the umpire, it does not mean that everyone will be happy with your decisions. But the rules were simple, ground threat calls stood. Those who disagreed too vehemently with the umpire become prime candidates to be the practice dummy for the next day's SAREX. Somebody had to sit on the desert floor to be 'rescued'. They were as good a pick as any. Because I got to select for the SAREX, crews tended to be nice to me, especially in the snack bar. Sometimes when a particular pilot was overly nice and generous, I suspected that his performance that day may have left something to be desired and he was worried that he showed up on the threat scoring logs.

I was never inundated with crew members who begged to miss flying one day so they could sit out in the desert while other flyers coordinated and implemented a rescue mission. Any time a pilot was out in the desert, we put a Para-rescue (PJ)/Survival specialist out there as well, staying far enough away from the rescuee so as not to interfere with things, but close enough to make sure nothing untoward happened, or if it did medical help was at hand. We had one para-rescue sergeant who was a giant. "Grubs" was known to carry a 60 lb pack with the standard rescue equipment and supplies, plus some other items. If he decided the trainee was doing well and it was a rare overnight SAREX, he would prepare what was often described as a five-star dinner out there in the sand and among the scrub.

While most crews begrudgingly put up with being picked, there was one particularly obnoxious major who tended to get on everyone's nerves. For some reason the fickle finger of fate pointed to him. He was not happy with the initial selection; but once up range decided all he should have to put up with was a helicopter ride to the range, a short visit to the desert, and trip back home. Not at the pickup site more than an hour, he began calling back to base requesting pickup. That was not scheduled until during the afternoon mission. He became more and more persistent and obnoxious; and finally way too demanding. That brought an Ops Cancel of the pick-up helicopter and he had to spend the night in the desert. His attitude had changed considerably by the next day.

Back in the saddle.

With luck, I could choke down a delicious greaseburger before heading back to Range Control for the afternoon 'go'. Same game, new name. It was pretty much a repeat of the morning, but the overall mission would be different. Each Red Flag was built on the idea of increasingly challenging and varying missions. What I loved to hear was how, as the days went by, it got increasingly hard for the Aggressors to score hits. The threat operators felt the same way. From their voices and comments they were increasingly pleased that their job became more and more challenging. They got a feel for who was dueling with them, and comments about a particularly slow learner that kept using the same unsuccessful techniques day after day could become harsh. These weak links often found themselves in a place of prominence in the next day's debriefing.

Pilots for whom there at first seemed to be no hope often showed improvement. Serious players would ask for one-on-one discussions about what they were doing wrong. They also asked for special attention to their tactics on upcoming missions. I would pass their call signs and planned tactics to the range, so operators could pay particular attention to them and give increased feedback.

The particularly hard workers sometimes earned "top learner" recognition from me during a briefing. More appreciated, was especially good comments from the operators. I tried to include that whenever possible during the second week of an exercise because what the threat operators said was far more valuable than anything I could say. Frequently, I was the recipient of their appreciation in the snack bar after the debriefing, always pointing out that the range rats were the real heroes.

Once the flyers heard this, they often arranged for a decent supply of 'liquid refreshment' to be sent up to the operators. Sometimes the truck bringing new videotapes back up range carried more than just videotapes.

Pilots who showed no improvement were singled out for attention as well. Some just did not get it. Others had an ego and attitude problem. While the Red Flag commander was in favor of singling out the improved pilots before the rest of the group, the un-stellar ones tended to end up in a private audience with the Red Flag Commander and his Squadron Commander. Usually appearing before the throne of the almighties did the trick.

On a few rare occasions particularly problematical cases were given airplane tickets or a seat on a supply plane back to home base.

The Daily Gong Show.

There was a standard lineup and pattern to the daily mass debrief that everyone had to attend. First up would be the Red Flag Commander and Operations Officer with words of wisdom about the plan, upcoming items of note, or other issues he felt needed to be brought before the group. Flying through a complete final approach with the drag chute deployed guaranteed a comment (it happened once), so did exceedingly gross behavior in the Officers' Club Flight Suit Bar downstairs (a place designed to keep raucous behavior away from the civilized patrons upstairs).

The briefing included safety warnings and on one occasion advising the participants to stay away from the Tailhook Convention going on in downtown Las Vegas; a timely alert, given how things turned out down there. Intel would re-cap the missions of the day and each participating squadron would brief their mission and results, as they saw them, with grease pencil slides presented by a designated briefer. The squadron brief told how well they felt they did, followed by operations briefers who sometimes told a rather different story,

Mass briefing   (Photo: U.S. Air Force)

The liveliest interactions came when the *Aggressors* made their presentation, fighter pilot to fighter pilot. Naturally everyone claimed to have shot everyone else down and everyone else figured they had been nowhere near shot down by anybody.

Some pilots came equipped with an orange Bulls**t flag to wave during these "discussions". The final result was usually settled with snippets of gun camera film. It was at this stage where who owed how many beers to whom in the snack bar was often decided.

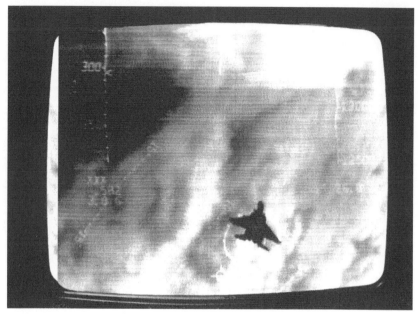

F-15 in Agressor's gun sights   (Photo: U.S. Air Force)

My Turn.

The final minutes were panic time. The mass debrief was already started and it was not going to wait for me to figure out something meaningful and worthwhile to say. Fortunately for my sanity, nothing that happened that day would be in my briefing.

The results from the previous day had been compiled and a complete, or nearly complete representative video done. All that was needed was to grease pencil in the Call Sign by Call Sign slides of yesterday's results, run through the videotape and come up with some useful comments, add in any particular learning opportunities gleaned from the overall mission results, include special notices about upcoming threat activities, and present it to a couple hundred aircrew who were tired, testy, and ready for cold beer.

Fortunately, I have seldom been at a loss for words. There were days I came screeching into the backstage area just in time for the stage manager to sling a microphone around my neck as the first slide popped up on the screen.

Starting with a summary of the previous day's events and scores, I had plastic slides which tallied the results as reported by the threat operators and validated by myself and the other threat analyst. It was also a place where any special information or lessons learned could be shared. Facing a room full of fighter pilots who did not like to be shot down made me nervous at first, until I realized how interested they always were in the ground threat part of Red Flag. While the *Aggressors* often got an argument, I usually got questions about the right way to survive.

It was interesting that while the air-to-air result briefings sometimes brought out a disagreement from the audience, there was very little pushback over the threat results. At first I thought this was because this non-flyer was critiquing a bunch of flyers who thought he had little to tell them. But it quickly became apparent that they were actually listening and absorbing what I had to say. That's the kind of thing that made my day.

A Bus Full of Range Rats   (Photo: U.S. Air Force)

The Story of "Fang"

From time to time I would pull a trick that always tended to knock the pilots out of their flight boots. The threat operators, as was explained before, were taken from one of the radar operator career fields and assigned to the range, doing a job not at all like what they were trained for – just like many of us at Red Flag. Not everyone adapted to the job or liked the working conditions. Others fell in love with the job, tolerated their living conditions, and became incredibly skilled at not only running the simulators but putting enormously valuable insights on the audio tracks of their tapes. They were also cool-headed and valuable during emergencies.

One such person was a young Airman Second Class who earned the tactical call sign "Fang" because of her skill and technique at the controls of a threat. She quickly got the hang of operating her simulator with, as far as the pilots were concerned, uncanny realism. The crews came to recognize the voices of some operators from the threat films. They got to hear her a lot because of the number of encounters with her "guns" and the importance of what she had to say to whoever she was shooting at.

Her voice comments were straight and to the point. At the start of a mission she had a sweet, level, lilting tone. This typically lasted through the first two or three engagements. Then things changed. She really got into it and her voice began to take on a new tone. It was a treat to listen to her audio and what happened next. Her voice began to get louder, harsher. Suddenly one of the attackers prompted what earned her the call sign. It was as though, like the sidewinders that kept them company up range, "Sfttt!", the fangs came out, and she snake-struck anyone who crossed her path in an airplane. She could provide the best threat operation narrative imaginable. The pilots loved hearing her deadly viper growl as she shredded attack formations, relishing every minute of it.

Fighter pilots, by nature, are competitive and don't take kindly to being constantly bested in a combat exercise; a good characteristic in that line of work. But when they are bested time and time again by a

Brunhilda-sounding threat operator, that did not always sit right. One of the themes at Red Flag was helping pilots learn to avoid falling victim to their emotions and jumping to conclusions that could cloud their tactical judgment. They had to develop the ability to deal with every attack in a realistic, not emotional way.

When pilots began to focus too much on not being hit by "that vicious Amazon" (or occasionally other uncomplimentary names), I had a ploy that tended to break their habit of letting macho fighter-jock assumptions get in the way of good tactics and planning. I started hearing plots planning to specifically teach "Fang" a lesson (by then I would have let the name drop). I could tell when pilots began focusing more on defeating her instead of the tactics needed for doing their job. It was time to implement 'Operation Fang.'

We would arrange for her to come down to Nellis for a couple of days R&R. I would have included my "don't jump to conclusions" speech in a couple of debriefs leading up to the big event. On the fateful day, I would select some threat video when she was at her growling, menacing, deadliest best. At that point I would announce that "Fang" was down from the range and would like to make a few comments about their performance. As the grumbling increased from the audience, she came on stage.

It may not have been totally fair to take advantage of over-sized macho egos (for purely educational purposes, not to burst some fighter-jock bubbles). She had gone over the top with makeup and clothes. Expecting a seven-foot, three hundred pound opera singer with horned helmet and armor, what they got was a petite, attractive young woman with long brown hair and a voice as sweet as honey. The next sound was that of a hundred jaws dropping open and from the stage it was quite a view, all those guys staring dumbly at the stage.

At this point, she would continue the speech about not jumping to what could be dangerous conclusions that drive bad tactical decisions, and remember "missiles and bullets had no gender." She would then give a PhD-level talk on how not to get blown out of the sky by AAA, especially the deadly ZSU-23/4

For a couple of days, some participants admitted that they got the point. Through the rest of the exercise "Fang" started having a hell of a time getting the attackers in her cross hairs or keeping them there long enough to score a kill. They were really doing it right. That always brought a big smile to her face that reflected satisfaction that her efforts and theatrics probably meant that a few more fighter pilots would bring improved survivability to any future fight.

The Show House

The Red Flag auditorium was designed with the technology of the day and size of the exercises. The Red Flag squadron plaque collection was wonderful. There was a memento from every unit that ever participated. They were mounted on the walls and added a certain air to the room. It was a most impressive record of U.S. and allied air forces and naval aviation units. Squadron stickers found their way to nearly every flat surface everywhere in the building.

A Red Flag debrief (Photo: U.S. Air Force)

One plaque could not be displayed, however. The Royal Saudi Air Force brought one to commemorate their first Red Flag participation It was magnificent – and made of gold. That got locked away in the Commander's safe.

We were not above a little humor, especially when it helped bring down the crews a peg or two. One trick we pulled occasionally during the first mass in-brief did double duty. It poked a hole or two in the sexist balloon of some flyers and woke them up during long, droning briefings on routes, call signs, frequencies, and other boring things. It also amused the hell out of us.

There was a tall, willowy blond from data entry who would wear a snake for a belt and wander the aisles of the auditorium during the opening briefing. It was a rather laid-back king snake who, for some reason seemed to find this a cool thing to do. Fighter pilots being fighter pilots, could not help but sneak a stare, getting the shock of their lives when her belt stuck its tongue out at them. Even from behind the stage we knew where in the auditorium the snake lady was just by listening for where the commotion was coming from.

The Threat Analysis office also managed Red Flag memento sales, especially unit participation t-shirts which proved the wearer's bragging rights for having participated in Red Flag XX-X. Especially treasured were the special made shirts listing multiple Red Flags. The shirts were individually designed for each exercise and squadron by a local retiree vet who did this mostly as a sideline. With a silk screen machine that he somehow inherited, Arky managed to provide a cherished remembrance for the pilots and a little money for himself, sometimes. He also was available to create special souvenirs for participating units.

Special Guests

VIP visitors were nothing new at Red Flag and could be a problem, getting in the way or taking up staff time. Generals, Admirals, and politicians dropped by regularly; often necessitating guided tours, private briefings, and select reserved seats in the auditorium.

Some, like the commander of a Wing whose squadron was participating, were to be expected. Officers from Pentagon tactics and planning shops, OK. A Supply Chief from Command HQ using it as an excuse for a trip to Las Vegas, why?

Generals, unless one or more of their units was involved, could be a pain. So could Members of Congress and their staff, although the occasional visitor from one of the Armed Services Committees was useful, especially leading up to budget time. The Red Flag Commander had to take time with them and the staff often had to put their work on hold to brief our guests. And they took up prime seats in the briefings.

Exercise participants were naturally told to "be on their best behavior" while the visitors were around; but this often put a damper on the better questions and discussion at the debrief. The guests usually listened patiently, had smiles and good words for the participants, then headed to the Officers Club where drinks were put on the CC's tab.

Sometimes the VIPs created a bit of a problem for me, like the time I was rushing down the hall for my threat briefing, rounded the corner at full throttle, and just barely avoided rear-ending the Air Force Chief of Staff.

Although it may just be fantasy-driven ego, most VIPs seemed to be fascinated by the threat range and expressed great respect for the crews and what they did. I, by association, seemed to be the recipient of a lot of the respect and curiosity. This had nothing to do with rank, especially from flyers who understood.

One Extra-Special VIP

One day, however, there was a special guest. Down the hall came an entourage with Sammy Davis, Jr., in tow. The Wing had brought him to the base for a tour that included a ride in the back seat of a fighter jet. As a result, he was wearing a loaner flight suit. What stands out in my memory was, because he was so short, the sleeves were rolled way up so his hands could stick out. The other memory was one of the warmest, most appreciative smiles I had ever seen. He was obviously having the time of his life. Showing Mr. Davis even the smallest thing drew a wave of "thank you's" and praise for the people.

At the end of the visit, the star of the famous Rat Pack said that in appreciation he would bring his show to the base for a free performance. He always sold out downtown in the casinos, so we figured it was a nice gesture and that was that.

Many months later, a notice went out inviting everyone on base to the Sammy Davis, Jr., Show on the flight line. The night arrived and the weather was frigid, with a biting cold wind blowing in from the mountains – a problem because in order to accommodate everyone the stage had to be set up in the open on the ramp by the maintenance hangers. No one would have said a thing if he had come out, sang a couple of songs, thanked us all, and headed back stage where it was warm. Not so. Sammy Davis, Jr., with all of his backup singers and dancers put on one of the best performances I'd ever seen, staying out in that freezing weather for nearly two hours. The only accommodation for the cold was skipping the normal intermission "so we did not have to sit out in the cold." What a guy. What a troupe of performers.

To the Silver Screen

Red Flag: The Ultimate Game This movie came to Nellis and worked with the base and staff to create a flick that, while well-intentioned, was mercifully short-lived. The story followed a pilot through an exercise and through all sorts of plots with rival pilots, ex-wives, etc. Fun movie, lame story.

I was away at Maple Flag during the filming and missed all the fun. Many of the Red Flag staff were used as extras. My flight suit had a prominent role costuming the Red Flag Orderly Room clerk's husband in a scene watching an ACMI display which simulated the threat range (which could not be shown due to classification restrictions). What he watched was a fictional briefing that filled in for me and the threat operators.

The movie crew was fascinated by the munitions handlers and their handiwork with things that go "Boom". They loved watching them prepare aircraft for flight.

The star was strapped in the back seat of an F-15 with a camera focused up on his face to get footage as the plane maneuvered. The shots were to be used in his flying scenes; but none of the footage was suitable for use. Either his face was in shadow, the sun was shining into the lens, or he was throwing up.

The World Premier was held at the Nellis base theater. Little was seen of the move after that. It may still be available on-line somewhere.

In an Officers' Club scene, many base people were used for the background party. At first, they gave out real drinks; but the stars kept flubbing their lines, making a new shoot necessary. After many re-takes, the Director switched to non-alcoholic drinks because the party scene was getting too real.

In the movie, they could always find a parking place anywhere on base, stretching credulity. It was fun to watch a car pull out from in front of the Thunderbirds' hanger and turn right, directly into a parking place in front of the 57th Wing HQ. The two places were actually on opposite ends of the flight line and called for a left turn from the parking lot. The roads and driving were way out of sync with reality.

Also, the Air Force does not typically hand-deliver travel orders to Staff Sergeants lounging by the pool at the Sahara Hotel and Casino – even if she is a beautiful blond and ex-wife of the leading character.

The New Building

When I left Red Flag for the Telecommunications Staff Officer Course at Keesler AFB, Mississippi, construction had begun on a major building expansion. When I got back, the Red Flag Headquarters building was twice as big. My new office was in the

Range Group building next door, where I became Chief Communications Project Engineer for the Nellis Range complex.

The renovations to the new Red Flag building essentially added another building behind the old HQ and a second floor. More seats, better comfort, and enhanced audio visual made the auditorium a great place for briefings. Back projection of the threat briefing slides and simulator video made that part far more professional.

This was before flat-screen TVs and computer-driven video. The threat slides were clear plastic and grease pencil. Three-color back projection TVs projected video recorded on commercial-size cassettes the size of a paperback book. All records were paper and pencil, although there was rudimentary computer storage of overall exercise records where clerks typed daily threat results into a data collection system; a far cry from what is standard today. PCs were scarce and limited. Three years later, after I had retired and was working for an ATC communications company, we would gather in the VP's office and bow down before his computer that had a whopping 5MB hard drive! It was considered miraculous at that time.

Working With Weasels

Back in the early to mid-1980s, the F-4G Wild Weasels from George AFB, California, were regular participants in Red Flag exercises. The idea was for squadrons to learn the best way to incorporate the unique capabilities of the Weasels into their missions. Naturally, in this environment, a certain level of competition and rivalry was bound to develop. Given the nature of fighter pilots and a few post-mission beers in the snack bar, "interesting" ideas were to be expected.

Enter "George, the Wild Weasel." Actually, George was a ferret, not a weasel. He was the pet of one of the squadron's maintenance sergeants who would bring the little guy to Nellis during deployments. If one of the participants was singled out for special attention, usually because of being found particularly unappreciative of the Wild Weasels or for hatching a really bone-headed mission plan, the NCO would bring her pet to the snack bar and let the furry little guy demonstrate his unique skill. Ferrets can squeeze into the tightest of spaces and have needle-sharp teeth. George found it quite amusing to scurry past the zipper at the bottom of the victim's flight suit, climb up his leg and – well you can figure out what happened next.

Weasels on the ramp. (Photo: U.S. Air Force)

Needless to say, the rest of the group found the recipient's reaction very, very amusing; and we were constantly amazed at George's ability to make an escape back down and scamper to safety before the victim could figure out what had happened.

After a couple of these Red Flag Wild Weasel George missions, the Commander decided to ban the little nipper from the snack bar before someone really got hurt. He also pointed out that getting miscreants by the nether regions so their minds and hearts would follow was a command prerogative.

As they say, those were the days. I sometimes found it hard to believe they were actually paying me to have so much fun. Ground threat analysis was an opportunity to "shoot down" airplanes by the dozen.

On a more mission-oriented note, I worked closely and repeatedly with the Wild Weasels' and their F-4Gs. Weasels used the range and threats to practice battling a ground threat while participants learned to incorporate defense suppression into their mission plans. Working with the Weasels was always exciting. I could not help but feel the joint uniqueness of our jobs.

This close work necessitated a few trips to George AFB, Victorville, California, to discuss the workings of the treats and coordinate with the aircrews. At one point, I was scheduled to fly in the back seat of an F-4G so I could experience a mission from their standpoint. The Air Force sent me to Edwards AFB for altitude and parachute training. A B-1A was parked close to where I was staying. A classmate got a tour of it through a buddy who was stationed there, but did not invite me because he thought I would not be interested in the only one of these historic bombers left.

The altitude chamber was interesting, to say the least. After some classroom work we were put in the chamber used in the past for astronauts. Sitting in a circle with our masks on, there was a big bang and the hatch opened sucking all the air out of our snug steel drum. It was then I learned that in this kind of situation the gas inside one's body wants to escape to balance air pressure between inside and out. But there are only two places where it can exit; there's a mask on one and you're sitting on the other

The ejection simulator gave me such a kick I the rear that I decided I'd rather not have to use it. The whole experience convinced me that some people have it right when they criticize sky diving, noting that it makes no sense to jump out of a perfectly good airplane.

The fateful day arrived. All suited up and looking like I knew what I was doing, I strode excitedly up to that hulking F-4G sitting on the George ramp. After climbing up the ladder, I accepted the help of a kind crew chief to strap in and hook up. I paid particular attention as he showed me where the injector handle was up on the canopy rail. I also very carefully listened as he told me under no circumstance should I touch it. Let the pilot decide if something happened that would necessitate us walking home.

After all that, I got all the way to strapping into the airplane trying calm my nerves, and being thoroughly briefed on the instruments, back seat displays, and how to work the intercom. I began to feel relatively comfortable sitting there strapped to a real fighter jet like an old pro for the first and last time.

Sitting there in my mask, helmet, and parachute, it was exciting to look out over the cockpit rail and see the crew chief twirl his hand in the signal to start engines. Then, even though the pilot and crew chief did their best; the damn plane would not start and I never had my chance to play fighter guy.

Special Demos

The Nellis Range has long been a place to develop and test a variety of weapons going back to WWII. As a result, someone decided it would make for wonderful, impressive VIP demos of some things. Close to the base, there was a wide flat area with a great spot for bleachers – and relatively easy access to/from the base and O-Club. While this seemed like a good idea, sometime things did not go quite as planned.

In one case, the Air Force had been evaluating a variety of ground-to-air weapons. One was a version of the Dutch Flycatcher anti-aircraft system. The idea was to take advantage of the facility to show the system to a VIP group of officers and contractors. The system was set up in full view of the bleachers, with the plan being to fly a target past the visitors, demonstrating how quickly the system could lock on and track a low-flying, close-in target. I was there to discuss target acquisition and tracking, should anyone have questions.

To accommodate the guests, the area featured a source for refreshments and special field facility for their comfort. Since they were VIPs, no regular "comfort station" would do. This one sported a solar-powered vent fan.

During one session, the crowd gathered, sitting in anticipation of watching the radar do its thing and show how quickly and effectively it could find and focus on targets from a variety of bearings. The event started and the radar was anxiously waiting to do its thing, search antenna twirling, targeting antenna quivering in anticipation.

One of the guests needed to use the facility, flipping the switch and activating the fan. For some reason, the targeting antenna decided that was just what it was waiting for, twitched a couple of times, and 'twang', zeroed in on the twirling blades, slewing the high-speed cannon directly to it. Fortunately, it was not a live-fire demo, so there was no harm done, except for the composure of the guest and the embarrassment of the demonstration director who nearly bagged a four-star general.

In another instance the demonstration was set up so a simulated airborne target would be 'flown' across the area on a suspended line simulating a hostile helicopter. A shoulder-mounted SAM would be fired at it to demonstrate how accurate the new missile was by blowing up the target. Naturally, planners wanted to have the test as impressive as possible for the audience, so a little explosive was added to the target to be set off by the demonstration team. Instead of a normal explosion, the idea was to show the missile blowing the target to smithereens.

Here's a hint. In a case like this, make sure the guy with the explosives trigger switch has his act together. For some reason, the target swept across in front of the VIPs, exploded impressively, and the missile then fired. This takes much of the credibility from the demo.

The Petting Zoo

When I was there, across the street from the Red Flag and Range Group buildings was a highly classified area we called the Petting Zoo. It took special access clearance to get in to view the displays, giving visiting aircrews a chance to see up close and touch a variety of Soviet ground threats and a couple of MiG aircraft hidden away inside a metal building. It included the opportunity to play with a real Soviet 57mm antiaircraft gun and get up close and personal with a nasty ZSU-23/4. There were lots of other trucks, guns, and other things to play with.

The place was surrounded by a high fence; but a few strange shapes stuck out over the top, raising the curiosity of the uninitiated and really getting crews in a lather for their chance to go inside for their scheduled visit. Units were usually given a scheduled date and time for their tour. It was like watching third-graders looking forward to their field trip to the zoo. Units that did attend were sworn not to tell those who had not what they saw. Unfortunately for the poor guys, most of the staff kept jabbing those who had not yet been about what they were missing.

A Mig To Play With   (Photo: USAF)

It was a wonderful opportunity for aircrews to get that close to the real thing, making anti-aircraft real and not just an abstract system symbol on their self-protection sensors.

In those days the Cold War was still on. Everything there was acquired in a variety of ways, often either bought from allies who had a stock of the stuff lying around or just plain stolen. Most of it no longer worked, so the prior owners could get some cash for what was to them junk; but which at Red Flag could be a valued training opportunity.

Today the place has been re-done with covered walks and is open to the public. The stuff there is no longer classified. The Petting Zoo has become a great place to bring kids and let them use old war tools as a jungle gym.

37 degrees right ... 44 degrees up ... FIRE!
(Photo: U.S. Air Force)

Interesting Playground   (Photo: USAF)

MGM Grand Fire

One morning, November 21, 1980, a devastating fire broke out in the casino of the MGM Grand Hotel and Casino in the heart of downtown Las Vegas. The fire spread rapidly through the first floor and a black, smoky cloud spread over downtown. The emergency response was massive.

Deadly blaze at the MGM Grand
(Photo: Las Vegas Fire Department)

Nellis had its regular contingent of helicopters and they immediately scrambled to assist. Red Flag had search-and-rescue scenarios planned, so there was a significant presence of visiting helicopters and crews. As soon as the fire hit the news, the Red Flag staff scrambled to the office. The Commander told the Director of Operations (DO) to put the word out, calling the rescue units to the base to take part in whatever rescue efforts they could. The DO said that calling them was not necessary. They were already pre-flighting their helicopters because as soon as the crews heard the news, everyone immediately headed to the base. So did just about every other Red Flag participant. Their only question was "what can we do?"

According to Las Vegas Metro Police, they requested "all available helicopters. Twenty responded, including nine USAF helos." 300 people were rescued from the roof of the casino. 5000 people were in the MGM when the fire broke out. 85 died and 700 were injured.

According to the Las Vegas Fire Department report, there was "a major problem with the use of helicopters involved, communications, rotor noise, and rotor wash." As the large helicopters (USAF MH-53s) hovered above the hotel, noise was so severe that fire officers had extreme difficulty hearing their fire radios. Rotor wash from 23 stories above, was severe enough to blow debris and blankets."

It ended up that the Air Force helicopters proved valuable as command and control platforms for fire and rescue. At one point the base helicopters helped direct air traffic over the scene, helping prevent further disaster because the smoke caused such severe visibility problems.

Bald Mountain Encounter

From the start, not "busting the box" and crossing the border of *Dreamland* was drummed into every pilot. Bald Mountain was a clear landmark on the northern border of the super-restricted area. As long as you flew on the north side everything was fine.

Life In Red Flag

This is way too close

(Photo:U.S. Air Force)

One Red Flag I was given an opportunity to fly with a three-ship of C-130s on a simulated parachute drop of supplies to a simulated ground party up range. I was flying in the second airplane. A visiting general was in the first C-130. As we entered the range, Bald Mountain was off to the left, just where it was supposed to be. Looking out the cockpit, I saw Number One sliding left toward the one spot on earth no unauthorized airplane should ever, ever, ever go. I had our pilot warn the other plane that they were heading for trouble. The answer came back that the general wanted to see what was there and it would be OK. Knowing better, I made sure we and Number Three passed comfortably north and out of danger.

We did the drop, a fun experience for me, and flew back to the base. Taxiing to our parking place, we shut down the engines. Number One was on our right. The propellers had not stopped turning when three security vehicles pulled up; not base Air Police but a far more serious bunch pulled up next to the crew hatch. Several nasty-looking, non-smiling guards in black stormed on board and hauled everyone, including the general, off to the vans. The general seemed to be trying to, unsuccessfully, make the case for the deviation being OK.

As the vans pulled away, my crew thanked me for the warning and we headed for the Red Flag building for a debrief and the snack bar. Meanwhile, the crew from Number One ended up in one of the hours-long, grueling security debriefs reserved for anyone who failed to heed the warnings; probably followed by a no-notice trip home, the general included.

Not flying into *Dreamland* is one warning about which no one was kidding. In my whole time at Red Flag, this was the only incident I knew of. Our warnings seemed to work for attendees. An unauthorized incursion could be career-ending.

International Participation

Over the years, Red Flag has become more and more an international exercise. During my years there, foreign participation was in its initial stages. The British and Australians were regular players, but I took part in the first time participation by pilots from France and Saudi Arabia. The inclusion of other air forces increased the international prestige of Red Flag. Being invited to participate became a badge of honor for air forces world wide.

The Brits were great team mates. During the initial team briefings they always emphasized the common background and partnership of the U.S. and U.K., "well except for that little thing back in 1776." Because they were regular participants, we had a smooth, productive working relationship. They were equal team members, once we got past the language barrier.

The French, on their first-time visit were one of the most enthusiastic groups I worked with. They came ready to rock-and-roll. The pilots were most excited by at the opportunity to work against the ground threats.

They were serious about going against our simulator operators. Some wanted to be able to make a run through the range on their assigned mission and then swing around and go back through again. Laudable, but an impractical idea. They also wanted to be able to use the range outside of Red Flag times. Not a bad idea, but the range was a busy place. We did manage to get a few of them a couple of extra passes, something they appreciated very much.

French Jaguars on the ramp. (Photo: U.S. Air Force)

There was one issue, however. They brought French treats for a party celebrating the FAF's first Red Flag; this included a massive amount of French champagne. Some really good stuff; but there was a problem. Some of the French officers brought their swords to the party. Holding the bottle up with one hand and swiping the cork with the sword in their other, they popped the cork neatly and cleanly.

In the beginning, this amazed the other pilots; but as the party advanced the amazement became envy then challenge. Many of the U.S. flyers decided that it was something they could do too. Drinking champagne did not do much for the Americans' acquisition of the French officers' skill, and instead of deftly popping the cork the U.S. pilots became adept at decapitating the bottle in a spray of champagne and glass. By that point, nobody much cared how the bottles got opened. It did make a hell of a mess, and the neck remnants all still had their corks in place.

The French were serious about getting the most from this training opportunity. They worked hard to maximize every minute of range time. On one mission, one of their planes was spotted trailing smoke and fire. Range Control immediately called a "knock-it-off" and the threats went into our emergency routine, with everyone focusing their cameras on the problem aircraft to insure a visual record of events.

The French commander got very upset with me for not engaging them during the emergency. He pointed out that they had come a very long way for this training and saw no need to cut anything short just because someone was on fire. Pointing out that one of his guys was in danger did little to reduce his anger.

This first-time participation followed me. Nearly two decades later, I met the squadron commander at an AOC (Association of Old Crows, the defense electronics trade association) lunch. I introduced myself and he immediately launched into an argument we had during the Red Flag about a kill score I made. Obviously, the Red Flag experience stayed with the French.

Australian F-111Cs came at least once a year while I was there. They were spirited players, serious fliers, and always vulnerable to our send-up of "confusing" them with the Brits. They usually got even during after-action parties. They could drink any unit under the table without even trying hard.

The first Royal Saudi Air Force participation was an adventure. Although there was an official squadron commander, a general, the real boss was Prince Bandar bin Sultan, member of the House of Saud, and later Ambassador to the United States. He was a pretty good pilot and quite the guy. His call sign was "Royal", naturally.

During one debriefing as Prince Bandar was explaining their mission plan, a smart-ass fighter pilot in the audience raised his hand and said that the slide was backwards (it was written in Farsi). The Prince took one look over his shoulder and, not missing a beat, replied "you're right," and had the slide turned around. I don't know if the pilot was right or lucky. The quick-thinking prince either admitted his mistake or just turned it around for the hell of it. None of us knew the difference.

The contingent planned a big party to celebrate the historic RSAF participation. The soirée was planned for the MGM Grand in downtown Las Vegas. Unfortunately, the fire forced other arrangements. The Saudis somehow got the Penthouse of the Las Vegas Hilton as an alternate venue. One usually had to book that room years in advance. Good food, some reportedly imported for the celebration, and expensive gifts for the staff; it was all there. All of the Saudis were very friendly and appreciative of the Red Flag staff's work on their behalf.

At one point, RF CC made the Saudi pilots turn and look out the picture windows. "Now that, gentlemen, is 300 feet (the range minimum altitude restriction for the exercise). This is probably the first time most of you have been this high in the last two weeks." This reflected their aggressive flying that tended to use the 300 ft restriction as a suggestion.

* * *

5. *Up Range*

66 **A**s far back as 1967, a group of visionary Air Force colonels had proposed 'Facility X', an electronic warfare test and training range. But the Air Force wasn't ready to invest the money and effort, and Facility X was never built. The concept returned with the *Red Baron Report*, and Facility X became the Continental Operations Range (COR), a very ambitious project that would have linked weapons ranges across the United States into a unified electronic warfare and bombing range." (*Red Flag: Air Combat for the '80s*, p. 59)

This proved overly ambitious at the time, but the Electronic Warfare Joint Test in 1972 proved that electronic warfare could be simulated realistically. Crews, until then, experienced threat strobes only in simulators or actual combat. In *Red Flag*, the author quoted the Range Group Director of Operations, "When Egypt and Israel went at it for those two weeks (in '73), we convinced ourselves once more that we had to be able to simulate the enemy's integrated air defense system and to learn how to defeat it or work around it, or we just weren't going to win. That's all there was to it." (p. 59)

In the mid-70s, the Range Group was given money and authority to reserve land, to purchase simulated threats and targets and to establish an organized range complex to support Red Flag and other combat training as well as electronic warfare testing. Planners established three range complexes; the Tonopah Electronic Warfare Range, Caliente Electronic Warfare Range, and Tolicha Peak Electronic Warfare Range. They covered about three million acres that were from 100 to 250 miles north of Nellis.

To control this airspace, radar inputs from FAA systems at Tonopah, Cedar City, and Angel's Peak were connected to a central Range Control System at Nellis. The Air Force also installed an Airport Surveillance Radar at Tolicha Peak for range use.

115

The various segments could be used separately, but for Red Flag they are grouped together to simulate invasion of a friendly country by a neighboring hostile nation. Live bombing could be done on the numbered ranges, although R-61, R-62, and R-63 west of the Sally Corridor and near Indian Springs were reserved for Fighter Weapons School and operational test and evaluation of new equipment. This included the Nellis Air Combat Maneuvering Instrumentation (ACMI) Range at Dogbone Lake.

Nellis Bombing and Gunnery Range

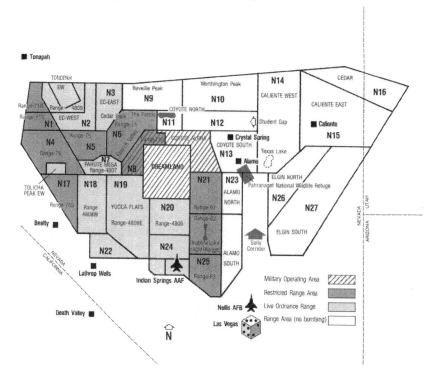

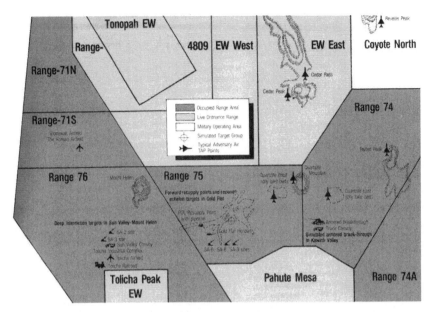

The EW Range Layout
(Source: Red Flag: Air Combat for the '80s)

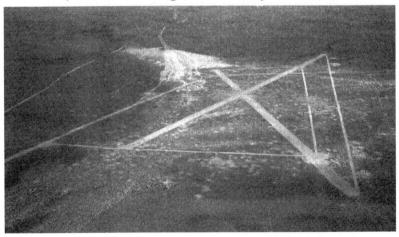

Airfield Target (Photo: U.S. Air Force)

Boneyard F-84Fs add realism   (Photo: U.S. Air Force)

SA-3 SAM site   (Photo: U.S. Air Force)

118

A truck park to beat up on (Photo: U.S. Air Force)

The 'seventy-series' ranges were where the Red Flag action took place. About fifty different targets are scattered throughout these ranges. They included simulated tanks, trucks, SAMs, AAA, airfields, helipads, industrial areas, bridges, radar sites, rail yards, trains, pipelines, and anything else the creative range folks could come up with.

Built up from all manner of cast-off equipment, telephone poles, cargo bins, dirt piles, plywood, and Styrofoam; the targets look pretty realistic from 500 feet at 500 miles an hour. They took quite a beating and maintenance crews are kept busy re-building all the stuff Red Flag pilots blew to smithereens on a regular basis.

Skinner gave a good description of the arrangement. R-74 was the easternmost range and arrayed as a "light ordnance area. "To USAF, which doesn't know the word excessive, 'light ordnance' means any bomb weighing fewer than 3,000 pounds. This is where they put any target they don't want bombed into oblivion, so this is where to look for the 200 Soviet tanks and armored vehicles simulating an armored breakthrough." (*Red Flag,* pg. 65)

R-74 was A-10 country, where the Warthogs live and work. It is the FLOT (Forward Line of Troops) and where pilots needed to watch for Smokey SAM.

R-75, further in was where newer SAMs lurked; SA6s and SA-8s could be found. A POL (Petroleum, Oil, and Lubricant) dump was protected by the T-3 gun-laying radars which controlled 57mm anti-aircraft guns.

The rest of the range included all manner of deep interdiction targets, such as airfields and industrial complexes along with their heavy SAM protection. Many targets could be attacked with live bombs. The threat simulators contained live people. Never the two should meet; so the sites were located in their own areas; EW East, EW West, Tolicha Peak, and Pahute Mesa. The ZSU-23/4s were sometimes trucked right up to the dividing line to make life interesting for attackers. Although the M-114s were mobile, they were usually carried to their site on a flatbed truck to prevent the damage that could result from letting them run amok across the desert.

The signal-producing simulators being offset by at least 5 miles from the target they are supposed to be protecting was not a major problem for most attackers. It could be an issue for the Wild Weasels accompanying strike packages. It was difficult for them to attack a radar site below their left wing while the signal is coming from far off to the right. This generated what is sometimes called "Red Flag schizophrenia". For this reason, the Wild Weasel squadrons did a lot of training independent of Red Flag when they could work directly against the simulators. This produced some very interesting battles.

To overcome some of this location problem, planners began developing the "nitnoi" ("no big thing" in Southeast Asia slang). This would be a small, low-cost unit with an antenna that could be located in the center of a target. A signal would be sent from a control unit and the little repeater would broadcast the threat signal as if it were the real thing and at the target, as it would be in the real world.

Before my tour was finished at Nellis, some of the early test units had begun to arrive. They featured an antenna that could be extended at the target and electronics box that could be buried or protected by a bunker. They were low-cost and easy to replace should a direct hit take one out.

Plans for further enhancements on this concept and other new range signal improvements were on-going. Time would see many technological advances make a big difference to range users in general and Red Flag specifically. One was the creation of a viable threat simulator industry that would grow and develop, taking advantage of new technology through the years.

Here I am, deal with it

I had to come to terms with my new assignment to the heart of the fighter world. It was necessary to become familiar with how I would come up with the information I would use to declare a plane shot down and why, and justify it to a pilot who most of the time considered it impossible that anyone could shoot down anyone as good as he was. There was not a lot of time do do it before my first Red Flag. I realized this could be a unique version of duck hunting in my native Louisiana – I shoot at them and they squawk. The difference, however, is that my trigger finger would be over a hundred miles long. Once I got to explore the range and meet the crews who would be doing the actual shooting, I was able to breathe a sigh of relief. The crews were good; damn good. This was going to be quite an experience.

My predecessor had shipped out before I got there, so there was no one to show me the ins and outs of being the Red Flag Head of Threat Analysis. Everyone would do as much as they could to help; but since Red Flag is the only place where this is done, there was no place else to serve as an example. To a great extent, I was on my own.

Here Are Your Faithful Steeds

One of the first things I got was an introduction to the Air Force blue Dodge crew cab truck. I would end up spending a lot of time with my butt firmly attached to the plastic-upholstered seats, seats that could really burn one's posterior when the desert sun was strong. This was a lot of the time, but could really be brutal in the desert summer. They were rugged, trustworthy steeds; some with four-wheel drive. This would come in handy when we had to access mountain sites around the high desert range.

One thing they were not, however, was comfortable. The springs were best suited for smooth roads with no load. The air conditioners, those that were installed, sort-of worked when they were really needed. The heaters could put out blast-furnace heat, especially when it was not needed. The motor pool was top notch, though, and really kept our trusty steeds running. They took well-deserved pride in keeping the fleet on the road. I cannot remember ever being stranded by one of the old girls.

A couple of interesting experiences involving me and a truck are worth noting. Once, an operator and I were pulling out on a road running along the edge of the range where we had been doing a special threat mission and got smacked by an 18-wheeler cresting a rise at warp speed. He spotted us pulling onto the road and hit the brakes, sliding past and clipping us with a rear tire. All our rugged blue tank got was a bent bumper. I ended up on the receiving end of a very pointed conversation from the Motor Pool sergeant for hurting his truck.

There was a wild mustang up range who seemed to have a bad opinion of blue Air Force trucks. Wild horses were commonplace up range. They were usually no problem and the range personnel, Airmen and NCOs, Air Force civilians, and contractors did what they could to protect them from starvation or thirst. But this guy really objected to a truck being in his territory, which he seemed to think was wherever he happened to be.

One day I was by myself doing survey work on the communications lines running along one of the roads between sites. I was at a cable head about 30 or 40 yards from the truck when I heard a horrid banging. There was this guy kicking the truck on all four sides. Job done, he snorted derisively at the truck and galloped off into the desert.

Trying to convince the motor pool sergeant what really happened was a challenge.

In the matter of the horses, there was concern about their starving or dying of thirst. Over population was an on-going problem. The Bureau of Land Management and range personnel did what they could to help the creatures. There was a way someone could adopt one of the wild horses and provide it a good home. A few young airmen thought having their own horse would be a good idea, especially some city kids whose knowledge of horses came from Westerns on the silver screen.

They coughed up 25 bucks and got a permit for one horse. The base gave them a rope and good wishes. The wild horses were not much into being adopted by some airman with a lasso, so this was fun to watch.

Getting There from Here

I had some trepidation over long up-range trips; but I soon began to look forward to them because it could be so exciting up there. The trips could be an adventure all their own. As they say, getting there can be half the fun. Some interesting things would take place along this route.

My first trip up was to learn about the range and the threats that would be the key to my life for the next six years. Immediately after leaving the base I turned onto Craig Road, passing the bar that would eventually host the Marine Corps Memorial mentioned earlier. In those days this was a great way to skirt the traffic and stoplights of Las Vegas to get to Highway 95 and head north, saving a lot of time and even more aggravation.

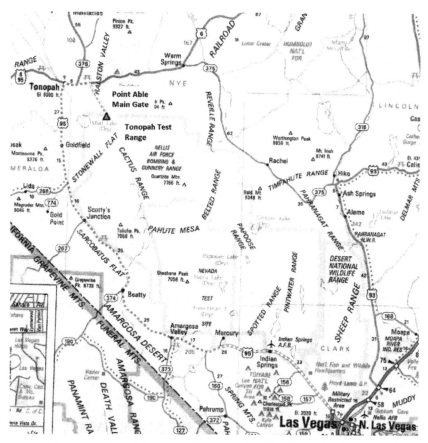

Except for occasional houses close to the Nellis end of the road and the (then) tiny North Las Vegas Airport, Craig Road was straight as an arrow, flat as a griddle, with the city off to the left and desert floor leading to mountains in the distance to the right. It was about 12 miles of dry sand and rock, mesquite, sage, and in the Spring a blaze of flowers. It could be especially beautiful when the beaver-tail cacti were in bloom.

Until my first range run, I pictured the area as dry, dusty, unremarkable. Halfway along the 12 miles to Highway 95, my head out the open window breathing in the desert air and smelling the desert plants, I began to like the place. Sections were very scenic with

124

desert bushes in the foreground and a low mountain range along the horizon that could be a color playground, depending on the height of the sun. It also turned out to be a favorite backdrop for filming car commercials for a while.

Today, Craig Road is a city street. What was open desert is now subdivisions and shopping centers, with a Beltway running north of the whole area. I'm not sure the change has been for the better. The same can be said of a lot of the Las Vegas sprawl today. After my time there, Vegas became like a second home town. But I'm not sure I would want to live there anymore.

At the end of Craig Road, I turned north on US 95, onto a four-lane freeway up a flat valley with mountains on either side. First came the small town of Indian Springs on one side and the (then) Indian Springs Auxiliary Airfield on the other. The first time through, Indian Springs was unremarkable and except for being an emergency landing strip and where the Thunderbirds practiced for their air shows. It seemed to be just a long runway beside the highway, with a few buildings scattered about. It was also where an unfortunate crash in 1983 killed four Thunderbird pilots when they flew into the ground – in perfect diamond formation. Today it is Creech Air Force Base, where most of the Air Force's Predator drones are controlled from.

Little did I realize that one day soon I would be called on to drive a very special trip up on a most unusual mission with eventual worldwide implications. One day, not long after I got there, the Red Flag Commander grabbed me and said that he had to get to Indian Springs NOW! Averaging over 90 mph, damn good for one of our old trucks, we got there in time for some high priority discussions and a conference call. As it turned out, the President had just approved and the Pentagon laid on a special mission. It was up to the Red Flag and Indian Springs people to figure out if ISAAF could be used to practice parts of a special operation. The answer turned out to be "yes".

This was the failed Iranian hostage rescue attempt that led to the *Desert One* disaster in April 1980 where a helicopter collided with a support C-130 during the attempt to refuel before flying to Tehran and rescuing 53 Americans being held hostage in the U.S. Embassy there.

Next came Mercury, the turnoff to the nation's nuclear test site; the only place on earth that repeatedly had atomic bombs dropped on it or blown up under it. I was a little excited the first time until I discovered that all I could see was a sign, an exit road, and a security gate. Like so much in that corner of the state, what you see is probably not there, or what's really going on.

The place would lead to some interesting visits. It can raise the hair on the back of your neck to stand at ground zero of one of the first atomic bomb tests or to go down nearly half a mile in the cavern at Yucca Mountain Nuclear Waste Depository (YMNWR) that was being investigated for nuclear waste disposal, a plan that has been the source of fighting for decades.

Mercury, the nation's nuclear test site.
(Photo: U.S. Atomic Energy Area)

Yuc
ca Flat, the nation's nuclear test site.
(Photos: DOE and Federation of American Scientists)

Bridge remnant from Ground Zero (Photo: DOE and FAS)

Entrance to the Yucca Mountain nuclear waste facility
Photo: Eureka County Nuclear Waste Office)

128

Once I was taken underground to where they were setting up for a bomb test, not a full-blown explosion. Underground high-yield testing had been stopped. I was standing next to an apparatus filled with wires and machinery while our host briefed us on the operation. There, peeking out from behind all the wire harnesses was a dull grey, cone-shaped object – the heart of everything. What an experience, to be casually standing so close to a real atomic bomb.

Up to Mercury it was four-lane highway; great driving. After a couple of runs it just became just one more interesting part of my new world. This is where the city haze gave way to the crystal blue sky Nevada was known for.

Protesting's not all it's cracked be

There would be adventures later; but one incident at the entrance to Mercury gave me a chuckle and I learned to take a lot of what went on in Nevada with a shrug.

Every so often nuclear protesters from one group or another would line up at the entrance with signs and jeer at the buses bringing employees to work, employees who generally ignored goings on at the side of the road. The only reason the riders were even awake was because they were approaching the gate and would have to show their IDs.

One particularly rag-tag group decided to jump the fence further up the road and penetrate the test site to show the world the wrongs of atomic bombs. Over the fence they went, running hell-bent for leather into the Mojave, cameras following them as they disappeared into the desert.

Site security ignored them; so a reporter covering the demonstration asked the weather-beaten old Chief of Security why weren't the guards going to chase after them and apprehend the trespassers? Typical of the practical approach to things around there, the Chief said, "Nah. With all the holes, snakes, sand, heat, and buzzards, they'll be back. They can't hurt anything but themselves in there."

Sure enough, about two days later the sorriest-looking, most bedraggled bunch you could imagine came crawling back over the fence to be picked up by some buddies who then headed back to Vegas. "Told ya," the Chief was quoted as saying. He had a point. The desert is no place for the uninitiated, as they found out the hard way. Luckily no one was hurt or died.

On the road again

From Mercury on, the run through the Amargosa Desert was a lovely stretch after I developed an appreciation for the beauty of Southern Nevada. After Mercury, the road necked down to two lanes and took on the unusual, interesting character of Southern Nevada. To the left was the road to Death Valley. Fifty years ago this trip would have been done by mule and wagon. Except for the asphalt, it probably looked pretty much the same.

Ahead was something out of the Wild West; without the silver screen cowboys. Ranchers still roamed the range on horseback. This would be a big part of my world, and quite a world it would be. Weathered wood, old buildings and rusty strands of barbed wire, worn-out wagons alongside the highway replaced the manicured lawns, paved parking lots, and buildings jammed together that I had lived with for so long. The rocky hillsides and ridges were gray, red, purple; colors that changed by the minute as the sun moved through the sky.

115 miles up the highway was the town of Beatty. It was nowhere near the size it is today. A couple of small motels where we booked rooms with vouchers from the Nellis Travel Office became a home away from home and whose owners became good friends. There was, naturally, some bars and a small casino that I really came to like. Over the bar was a sign noting that their tables and slot machines worked best with silver dollars, especially their own, "Because we like our money to go clunk!" Cold beer, good food, friendly staff; can't go wrong. The casino eventually added hotel rooms. It don't get much better than that.

The wonderful people made the town a comfortable stop for travelers and a second home for us and the range rats that worked at the Tolicha Peak EW Range. The locals knew what we did up there, not the restricted details (except for the civilians who called Beatty home); but they felt it when we were tired, were sorry when things went wrong, and celebrated when the Tolicha Peak range rats marked the end of a particularly successful or important mission.

Like many places in Southern Nevada, the area around Beatty was filled with old silver mines. It was a great place to explore, especially the ghost town of Rhyolite. There were the ruins of a former part of the town, with partial stone walls from an old hotel and other collapsed buildings.

Rhyolite, Nevada (Photos: Rhyolite.com)

131

It was interesting to walk around the hillsides when I had a little spare time. I came across holes in the ground with tin cans on a stick next to them. At first, I thought these were remnants from the sourdough days, but I found out that that even today those claims are legitimate. I sometimes would toss a stone down the mouth of one of these mines. It was fascinating, listening a long time to hear it hit the bottom. The silver days must have been very, very interesting.

This is also where I discovered the value of peacocks as security systems. A couple of old guys had a pair running around their cabins, which were surrounded by all manner of mining equipment, some of it museum quality. They also had a pair of large, scruffy watch hounds. Startled at first when I came upon them, I quickly discovered the secret was to walk quietly past so as not to wake the trusty guardians. But no one got anywhere close without those birds raising one hell of a ruckus. No sneaking up on them.

Getting to Tolicha Peak Electronic Combat Range was easy. Drive out of town to the north. A little past Fran's Star Ranch (not the kind where they raise cows) with the crashed airplane out back, turn right onto an unmarked road with the dilapidated windmill, and drive into the hills.

Fran's Star Ranch 2009. Hasn't changed much.
(Photo: brothelangelsladies.com)

Fran's Unique Road marker for Tolicha Peak Turnoff
(Photo: Nevada Aviation Hall of Fame)

When things were dry, which was most of the time, a truck kicked up quite a dust cloud. There was a running joke that if anyone stopped to relieve themselves on the side of the road, a tree would be sprouting there on the way back.

After Beatty, it was 93 miles to Tonopah, with the entrance to the north range 30 miles east of town. The desert beauty continued along this part of the trip. The flatness of the road was interrupted by dry washes from time to time; but variation began to sneak in. Some found this boring, I never did.

Further north the ridges came closer to the road and a few spots had US 95 curving along the side of cliffs with deep valleys down below. One impressive spot had the road taking a left (heading north). There was a deep valley on the right and suddenly popping into view was an old shaft house and derrick, weathered and rusty, still standing guard over an old mine.

Remnant of Old Silver Mine (Photo: Nevada Dept of Tourism)

I could never drive past without a surge of curiosity. What was it like here when the winch was working and sourdough miners were pulling bright treasure from the ground? One time I stopped to get a closer look. Expecting the curious to do just this, the state had fenced it off. Because the structure was perched on the edge of a steep cliff and over a deep mine shaft, keeping people away was a good idea. The curve was sharp and there was no place to park, another reason for on-the-move viewing only.

Home Sweet Second Home

From Beatty to Tonopah and the northern range desert travel continued, passing the marvelous little town of Goldfield, a few miles before Tonopah. It was a fascinating spot along the way and typical of what made these commutes so fascinating.

Goldfield, Nevada   (Photo: Goldfield Historical Society)

134

Tonopah was anchored by its Mizpah Hotel, a remnant of the old west. Built in 1907, the place looks older. This was the result of careful restoration. I stayed there once instead of using one of the regular motels around town. The place was still back in the Wild West; I expected to find Wyatt Earp's boots under the bed.

The Mizpah Hotel (Photo: mizpahhotel.net)

135

Unlike quiet Beatty, Tonopah was a major stop for tourists heading from Las Vegas to northern Nevada, especially Reno. As a result, it was more like towns all over. Utilitarian Tonopah did not have the charm and fascination for me that Beatty did; nor did I feel as much at home as there.

In Beatty and Tonopah the locals were unique, friendly people who respected what we did and were glad to have us around. Staying was always a pleasure since most of the time spent in town was for food, drinks, and sometimes a quick pass at the slot machines.

Everybody had their own story. Sandy, a waitress at one of the restaurants we frequented, kept a tally of the field mice her kitty, whose name apparently was "That Damn Cat", brought inside her house from the feed storage room and presented as a gift. They were knocked out, coming to and beginning to scamper about scaring the hell out of Sandy just as she came into the room. She never could break the stupid cat of such an annoying game. Every trip brought an update. These were the kind of experiences that made these trips so pleasant.

Tonopah EC Range

A short drive out of town and I would head east out of the crowds on State Highway 6, driving 11 miles to the Sandia Rocket turnoff and in 30 miles to the Point Able Main Gate, the entry into the surreal world of the Tonopah Test Range and the northern EW ranges.

Hidden in the desert between Tonopah and the northern range border was a nearly invisible remnant of a runway where pilots landed while conducting a variety of World War II training and test work. A few vestiges of hangers and other temporary buildings could be found if one knew where to look. It is still marked on a few older aeronautical charts.

The two-lane access road ran through an interesting landscape of mesquite bushes, bombs, and rockets sticking out of the ground. The mesquite was normal, most of the bombs and rockets were the result of the area's history as a test site starting back in World War II.

Tonopah Test Range Turnoff (Photo: Bing.com)

There were many things dropped on or rocketed into this part of the landscape. The ones that could be seen were, supposedly, no longer live. In-briefings for new people to the range warned that sometimes after wet weather things would work up from below the surface. These could be live, the briefing noted, and drivers were instructed to carefully go around anything new sticking out of the road or range surface.

Reportedly a few past drivers ignored this warning and had a wheel blown off their truck. I think I was a little disappointed at never experiencing one of these sudden apparitions.

Reportedly, there was a wolverine that made its home in a culvert running under the access road between the gate and range. Word from the motor pool was that if anyone had a flat tire anywhere around the middle point, do not stop and get out so as not to get caught crosswise with "the mean little bastard."

Rumor had it that once a truck stopped and the wolverine bit a hole in the tire. While there was some doubt as to the veracity of the story, no one wanted to test it out. I never had a flat along there or an opportunity to find out.

The Back, Crazy Way

There was another route to the northern range. It was closer to a back entrance on the northeast corner of the range; but then there was a long run on a dirt road to get to our destination. It avoided traffic trouble on the main route; but more commonly it was an efficient way to get to some of the range communications facilities in the eastern part of the ground range. It proved useful while working on the microwave system installation a few years later. It was also an interesting route when escorting visitors, such as the contractors who would be building a new digital communications backbone. Some key microwave nodes were located on peaks in that part of the range.

This run involved heading northeast from Las Vegas on Interstate 15, turning north on State Highway 93 and up along the eastern border of the range. It ran through the beautiful Desert National Wildlife Refuge, the Pahranagat National Wildlife Refuge, and past an area of pleasant camp sites along a small river which, in spite of the desert, was usually flowing. A quiet, smooth run.

At the turnoff from Highway 93 there was a tiny desert truck stop with a large parking area, a couple of fuel pumps, and a cafe featuring 'the largest flapjacks anywhere'. It became a must-stop whenever I had to bring an outside visitor up to the range. They never got over the experience of the waitress delivering their order with the enormous flapjacks overflowing the plate and dragging on the table. They were unusual, and real good.

Once they'd been there, visitors talked more about the breakfast they had than the work they did. Anyone who had not been there thought I was pulling their leg. One stop and they were part of the experienced flapjack breakfast team. A couple of times, while arranging for a visit, the request was to go the route where the big flapjacks were.

In 93 miles, just past Alamo and Ash Springs, there was a turn to the west on Highway 375, listed as the Extraterrestrial Highway by the state of Nevada.

ET was never here (Photo: Nevada State Highway Commission)

This came from the odd assortment of UFO believers who, as a result of wild rumors that the government was storing space aliens from a crashed UFO in Area 51, the test area that "did not exist" just over the mountains to the west. While the real Area 51 was and continues to be an isolated place to do aircraft and electronics testing, the imaginary Area 51, where the aliens are stored and all sorts of interesting science fiction operations took place was far more fun to talk about.

Many scruffy wack-jobs flocked to the area, holding a few major gatherings through the year. Several more rational types found it profitable to run places that sell beer and hamburgers and cheesy souvenirs to the ultra-active imagination bunch.

139

A popular stop along this route was the *111 Cafe*. It was most popular on the way back after working up range. The name came from the owner's ability to expand the place with money from claims to the Air Force for damage caused by low-flying F-111s going Mach 1 too soon, before entering the range. The attraction was the cook who had an actual hook in place of a lost hand, something right out of pirate stories. It was worth the price of a hamburger basket to watch him spear a potato from the bin with the hook and in a blur of the knife in his one good hand produce what would soon become french fries.

One of the Stops on the Back Route   (Photo: Bing.com)

A little past the town of Rachel, the turnoff into the range was a much less dramatic entrance than coming in from Tonopah. Rachel advertises prides itself as the only town on the Extraterrestrial Highway, "population humans 48, aliens???" the road sign said. It takes advantage of being the closest town to Area 51, south of Bald Mountain.

I had little time for the scruffies. Besides, I would have hated to burst their bubble and reveal (even if I could without spending the next several decades in federal prison) that what really went on up there was actually quite boring. All science stuff. Their crazy version was a lot more fun and made for better stories. It would also destroy the market for all the made-in-China plastic aliens and flying saucers the proprietors sold to visitors.

Bald Mountain, at the edge of their fantasy land
(Photo: ThePost 2013)

Surprise! There is an Area 51

While I was writing this part, the CIA, through the George Washington University Security Archives program, released a 407 page report on the U-2 and OXCART (SR-71) programs. The report was titled *The Central Intelligence Agency and Overhead Reconnaissance: The U-2 and OXCART Programs, 1954 – 1974, approved for release 2013/06/25.* Most of this had already been published in Annie Jacobson's book *Area 51,* which was written based on information declassified by Leon Panetta when he was CIA

Director. She included a map of Area 51, or The Ranch as it was once called.

Naturally, there was a temporary spurt of chatter in some newspapers about this 'revelation'; including maps, which ran from fairly inaccurate to downright wrong. One reporter for a major paper wrote about how he had, as a result of studying the CIA report, "found the gate to Area 51 nine miles up Back Gate Road from just outside Rachel, Nevada." Sounds like what he found was the back gate I used to get to the northeast corner of the range. I hate to disappoint him and his readers; but there is no Area 51 access from there.

Adam Nagourney, in the 23 August *New York Times* wrote of his adventure and how he decided not to risk getting shot for violating the warning signs. In that area he was more at risk from rattlesnakes and dehydration than homicidal security guards. Those guys were elsewhere, and anyone who stumbled across them were left with no doubt they were in the wrong place and definitely not welcome. To give you an idea, while the official "No Trespassing" signs stated "Use of Deadly Force Authorized", that bunch painted over "Authorized" and scribbled in "Encouraged" or sometimes "Required."

Signs making it obvious visitors were not welcome
(Photo: Jonathan Strickland, How Stuff Works)

Adam's article described the "A'le'Inn" (pronounce it as it's spelled), a fun and funky stop for decent food, as long as you stuck to the hamburgers) and cold beer. I noticed one thing. From the picture accompanying the article, they haven't redecorated the place since I was running around up there. Wondering what effect this CIA announcement would have on the alien-seeking cuckoos, he asked Annie Jacobsen. She was right on target, I think, saying that this revelation "won't dampen interest in what lies behind the fences. It will only make people more curious."

Interestingly, at no time did any of the 'alien/UFO/the government's hiding something' bunch show any interest in what I or the range crews were doing. My guess is they sensed that whatever it was, it was not as exciting as little green guys from outer space. Besides, they and reality seemed to just be vaguely aware of one another's presence. Had a real alien plopped itself on the hood of their truck, this bunch they would probably immediately begin to believe in something else that didn't exist. This could be one creative crowd, maybe because of the funny little cigarettes they smoked.

Reality Strikes – They never had a clue

Interestingly, at no point was there ever was any indication that the dead-enders had a clue about what really was going on, especially at TTR (Tonopah Test Range). While the crazy fringe was focusing on how secret the government was keeping their stock of captured aliens, in reality there was a truly interesting and useful program right under their noses. Annie Jacobsen, in *Area 51: An Uncensored History of America's Top Secret Military Base* published the true story.

Jacobsen called the F-117 *Nighthawk* stealth bomber "the single most important invention in airpower since the Army started its aeronautical division in 1907." (pg. 339)

Beginning in 1976, development of the F-117 began at the Lockheed Skunk Works and eventually moved to Area 51 for flight testing. "But there was simply not enough flat square footage at Groom Lake to drop bombs. There was also the issue of sound. With multiple projects going on at Area 51, not everyone was cleared for the F-117. ... A second site was needed. (pg. 341)

"The Air Force turned to the Department of Energy, formerly the Atomic Energy Commission. A land-use deal was struck allowing the Air Force to use a preexisting, little known bombing range that the Atomic Energy Commission had quietly been using for decades. It was deep in the desert, within the Connecticut-size Nevada Test and Training Range. Located seventy miles northwest of Area 51, the Tonopah Test Range was almost in Death Valley and had been in use by Sandia Laboratories since 1957. The Department of Energy had no trouble carving a top secret portion out of the 624-square mile range for the Air Force's new bomber project. To be kept entirely off the books, the secondary black site was named Area 52. Like Area 51, Area 52 has never been officially acknowledged." (pg. 341-342)

Jacobsen noted that Sandia National Laboratories' dropped 680 bombs and launched 555 rockets at Outpost Tonopah between 1957 and 1964. This is the likely source for most of the strange things we would come across up there.

"Tonopah was so far removed from the already removed and restricted sites at Area 51 and the Nevada Test Site that no one outside a need-to-know had ever even heard of it.."

Construction for an F-117 support facility began in October of 1979. This included runways and taxiways, a maintenance hangar, and other buildings.

"Air Force officers assigned to the base were ordered to grow their hair long and to grow beards. Sporting a hippie look, as opposed to a military look, was less likely to draw unwanted attention to a highly classified project cropping up in the outer reaches of the Nevada Test Site. That way, the men could do necessary business in the town of Tonopah. ... Area 51 and Area 52 worked in tandem to get the F-117 battle-ready." (pg. 343)

The airfield at TTR   (Photo: Bing.com)

A series of screw-ups nearly outed the effort. This included a mock attack at the gate by security guards that nearly blew the project's cover. A foolish act by Lieutenant General Robert M. Bond occurred when he decided to fly one of the MiG aircraft in the Air Force fleet of Soviet-made aircraft kept at the site.

"General Bond requested to fly the MiG-23. There was some debate about whether the General should be allowed to fly. ... Usually a pilot would train for at least two weeks before flying a MiG. ... Instead, General Bond got a briefing while sitting inside the plane saying 'do this, do that'. In other words, instead of undergoing two weeks of training, General Bond pulled rank." (pg.344)

The general ran into an emergency "just as he crossed over into the Nevada Test Site. Bond radioed the tower on an emergency channel. 'I'm out of control,' General Bond said in distress. The MiG was going approximately Mach 2.5. 'I've got to get out, I'm out of control' were the last words. The MiG had gone into a spin and was on its way down. Bond ejected from the airplane but was apparently killed when his helmet strap broke his neck.

"The general and the airplane crashed into Area 25 at Jackass Flats, where the land was still highly contaminated from the secret NERVA (unsuccessful nuclear thermal rocket) tests that had gone on there (years before). General Bond's death opened the possible exposure of five secret programs and facilities, including the MiG program, the F-117 program, Area 51, Area 52, and the nuclear reactor explosions at Jackass Flats." (pg. 345)

According to Jacobsen, the Air Force decided to avoid detailed questions, "The Pentagon made the decision to out the MiG." this entailed a variety of stories about where the Soviet aircraft came from.

"With this partial cover, the secrets of Area 51, Area 52, Area 25, and the F-117 were safe. It would be another four years before the public had any idea the F-117 *Nighthawk* existed. In November of 1988, a grainy image of the arrowhead-shaped, futuristic-looking aircraft was released to an awestruck public despite the fact that variations of the F-117 has been flying in Area 51 and Area 52 for eleven years. The 'Wobblin' Goblin' as they came to be known came to their own publicly years later when they became the first stealth aircraft to launch several bombing attacks in several campaigns.

This being said, it proved interesting to operate an EW range right next to this place. As the one person who monitored all of the video from up range, I was a responsible for making sure no strange airplanes, or strange anything else, showed up on any threat video or at a debriefing.

The pilots flying Red Flag were too busy to notice, although once and a while someone expressed the hope that they would stumble across one of the MiGs and go at it with the Soviet jet. The goofballs along the Extraterrestrial Highway were too busy obsessing over aliens to care about any of it. Go figure.

Home on the Range

One skill necessary for entry to this side of the range was the ability to work a cowboy gate latch; a bent-wire hook that could be undone and re-done with one hand (designed for while sitting on a horse).

There were several cattle guards along the early sections of the range road. This was an area where free-range cattle tended to wander. I enjoyed driving slowly by, leaning on the window sill of the truck and looking at the cows looking at me. I wonder what they were thinking.

There was little to nothing happening on the ground out here, but Air Force officials would rather the cows stayed on their own side of the fence and reportedly kept a real-live cowboy on retainer to round up any four-footed hamburger factories who wandered through a bad spot in the fence. There was a rumor that it was a requirement that local ranchers had to brief their cattle not to go exploring on the Air Force side of the fence. The scarce grass was pretty much the same on either side, so why the attraction is anyone's guess. That's cows for you.

The range road was not bad, straight as an arrow. Range maintenance usually kept the surface fairly smooth. It was only after a rain that ruts developed where water ran across the road at dry washes. Because rain was infrequent I usually pushed the old range trucks way faster that I should have to get to our destination faster. Usually it was mostly a dusty, bone-rattling run.

But one day I was taking a couple of visitors to do some prep surveys for the upcoming communications job. Ahead, just above the middle of the road was a black spot coming at us. As we closed, it became a blob with wings. I was driving fast, it was flying fast, and we were on a collision course. Figuring the bird would zip over the top of the truck as we passed we kept going at speed, figuring we would get a close, quick look at some sort of raptor as we drove under it.

To our surprise, just as we came together the big bird dove at the truck. Either we nailed it or it nailed us, either way the result was a loud bang, an explosion of feathers, and through the rear view mirror I could see the bird flipping end-over-end in our dust trail. We knew we it would not do any good, but stopped anyhow and went back to see what decided to attack an Air Force blue truck. It turned out to be a large, heavy hawk. There was no way he could tell us why he thought attacking us was a good idea and what he hoped to accomplish by the last-second dive. Even if we were not closing at probably a combined 75+ miles per hour, he could not have carried us away.

Getting back in the truck, we saw the bloody feathers stuck all over top of the windshield and realized how close a call it was. Had he hit a little bit lower in the middle of the windshield Harry the Hawk probably would have crashed through the windshield and all of us would have been in sad shape, or worse. Life up range always had a way of throwing new curves. Always something different.

Tolicha Peak

The Tolicha Peak Electronic Combat Range (TPECR) was an important part of Red Flag and supported many other Nellis range and training missions. Located on Pahute Mesa, north of Beatty and west of Quartz mountain, TPECR was about 20 miles south of the Tonopah Test Range's southern border. When I was there the facility was much smaller than today.

***The Tolicha Peak Electronic Combat Range
(Photo: otherhand.org)***

Because it was closer to Nellis than Tonopah and getting there was a much more direct flight, it was easier and quicker to get to and from. Flights did not have to contend with the same air traffic issues as they did going the long way around to the north ranges and operations would not conflict with larger operations at TTR.

One system Red Flag participants made some use of was TOSS, the Television Ordnance Scoring System. Several targets were laid out on the range floor to gave pilots a realistic image to attack. Television cameras mounted around the area would record images of and triangulate on hits by 25 lb practice bombs. These Smokey Joes released a puff of smoke on impact.

149

The video showed the hit and operators produced a scoring sheet that could be used to debrief the attack. Sometimes there was a straight line of Xs running across the aim point, sometimes there was an X on the aim point, and sometimes Xs were scattered all around the area. Based on my experience with TOSS, in later years I got quite a chuckle when I saw in the window of an Air Force recruiter's office their latest motto, "No one comes close." I doubt they realized how true that sometimes could be.

The threat array at Tolicha Peak was less dense than at Tonopah and tended to work ingress routes to the target. They could work in conjunction with TOSS to determine the effect operating countermeasures had on a pilot's attack accuracy.

This is where a range control radar was located and where a key node of the future digital microwave system would be installed. ACMI, the Air Combat Maneuvering System here would eventually give birth to the Red Flag Measurement and Debriefing System (RFMDS).

Tolicha Peak was a great place for special electronic warfare evaluation missions. With its advanced electronics array, Tolicha was particularly well suited to EW data collection that could operationally validate system and/or technique effectiveness. It was also easier to provide access to allies and contractors for on-the-ground participation in things like measuring the effectiveness of new EW kit for just-purchased fighter jets by a valued ally.

We also did some interesting plane-to-plane comparisons of EW effectiveness. One of these efforts led me to recommend that the Air Force replace the EF-111A *Aardvark* with the EA-6B *Prowler*, which I found far more effective for escorting and protecting strike packages. No one took me up on the idea then; but in conversations with electronic warfare specialists in later years I was told my idea was not all that far off base. The Air Force retired their *Aardvark* fleet and began an exchange pilot program with the Navy's EA-6B squadrons.

It was also where range rats honed their skills at celebrating a successful project when all that could be said about it was "things went great." There were some very special accomplishments that gave rise to some very special parties.

One quiet day I was at the Tolicha Peak radar site. It was always thrill to look down into the valley with its targets and other facilities. There was a beautiful, crystal clear blue sky and not much going on. An A-10 was circling lazily over a bulls eye on the desert floor. He rolled to the vertical and smoke streamed out from the GAU-8 Gatling Gun. Seconds later the growling 'elephant fart' that the gun was noted for rippled through the silence. Then nothing. Another delay and suddenly the ground at the target completely exploded as the shells decimated the target. What fun to watch.

A-10 on the attack   (USAF photo)

Another A-10 story involved one inbound from the West to participate in an exercise. The hot desert air was causing the notoriously under-powered plane trouble. The pilot, seeing he would not make it over an upcoming ridge line punched out. Apparently, the loss of weight as the pilot and ejection seat went their own way allowed the plane to gain enough altitude to clear the ridge instead of auguring in. It apparently cleared another two before crashing. We wondered why the A-10 did not just keep on and land at Nellis like it was supposed to.

Water in a waterless place

One evening, a group of us had finished our work and were ready to head back to Nellis. After a bite at one of our favorite Beatty watering holes, we piled into the truck and headed toward home. At the edge of town a Nevada Highway Patrol Officer stopped us and directed us to turn back.

Although the weather had been fine in our area, a severe thunderstorm hit miles away in the desert and the dry wash south of town was flooded. Highway 95 was impassable and would stay that way for a while. "You may as well head back to town. No one's going anywhere tonight." A pickup truck had been washed away in the flood.

It was back at the cafe that I learned the Nevada way with beer. The bar tender set two on the bar; one to wash the dust out, the second to taste.

The next morning we could get through and drive back to Nellis. We found out that the truck was found five miles down the wash. Sadly, the driver did not survive.

In a similar flash flood incident back at Nellis, one time I went home to base housing for lunch. A storm hit in the hills to the northeast, a flood came down the middle of Nellis boulevard that ran past the front gate, and made getting back to work impossible. I got an unexpected afternoon off. Such was the way of the desert.

There I Was

On my first trip to the desert range, the rocks, sand, scrub, and mountains made me feel that I was in a separate part of a special world. It was a unique place, a unique job, and unique people; and here I was, right in the middle of it.

It would not be long before my first Red Flag began, so I had to hustle and learn how things worked, what the systems could and could not do, and how I fit into the whole thing. I had to trust the operators and they had to trust me; and we all had to play together to benefit the Red Flag participants. In addition, we would be called on to grow the range and find ways to make Red Flag better.

Although up range would seem strange at first, during the Red Flag exercises, with few exceptions, I had to be at Nellis. But I would be an extension of what went on up here. This place had to become part of me, and vice versa. With my first exercise coming up, this had to be more than just a sight-seeing trip. It was crucial, I felt, to instinctively feel what was going on in a threat simulator and be able to convey that in my briefings.

With my background of coming up through the ranks, I knew how counterproductive it would be to interfere with the airmen and sergeants. They knew what they were doing, so the secret to success would be to let them have at it. I knew instinctively that they would not lead the aircrews astray. The operators agreed and did a fantastic job bringing me up to speed fast and making me an extension of their efforts. Fom the start, we made a great team.

Up range, the facilities were not beautiful; but they were very functional. The EW range and targets were scattered across a flat desert floor ringed by red mountains in the distance, mountains that changed minute-by-minute as the sun moved across the sky. The dry desert sand was spotted with mesquite and sage brush. I quickly discovered that the desert sand was filled with tiny desert blooms called "belly flowers" because to truly enjoy their beauty you had to look at them while laying on your belly.

Walking around the hard sand and small rocks, it sounded like the a scene from the "crunchy-gravel mystery shows" of British television. At first glance, the place looked desiccated and dead. Looking around, I discovered it was filled with small creatures scurrying around, especially after the sun went down. Mice, kangaroo rats, lizards, small birds, insects, and an occasional sidewinder or two; they were all there. There could be a cow or wild horse wandering around as well. That, plus our own 'range rats'; and the place was alive.

Driving the range during quiet hours could be an almost religious experience; especially in the morning before the heat, as sunrise reflected off the mountains. Some visitors said the space was a wasteland. I felt sorry that they could not see and feel the beauty that I and the other 'range rats' did.

The weather was unique, severe clear or Clear Air, Visibility Unlimited (CAVU) most of the time. In the summer, days could be blistering 100s with the temperatures plummeting to the 40s once the sun went down. In the winter, the days were warm and pleasant, but night became frigid. During quiet times the smell of dust and mesquite permeated the air. The night would carry the sweet aroma of desert flowers and lights from miles away carried across the range.

During operations, the roar of jets and smell of burned jet fuel was everywhere, adding excitement.

Building a tool box

I arrived at the peak of "do-it-yourself" threat development season. For years, engineers had been using incredible ingenuity and a lot of old parts to create what, from the cockpit, was as realistic as could be. This was the result of years expanding Red Flag and appreciating the value of the ground threats to the training.

But it was not just the hardware. Threat operators had developed a confidence in and appreciation for the uniqueness of what they did. They understood the value of the training they provided. Pilots, should they ever be sent into combat, would live because of them. The Airmen, NCOs, and civilians were doing a remarkable job. I could hear it in their voices from the very beginning. They wanted to do the best job possible, and almost always achieved that goal, under some trying circumstances at times.

Everyone appreciated the value of these folks. The Air Force usually gave extended assignments to those who wanted to stay. Their unique skills were too hard to replace. A large portion of the workforce was provided by a contractor who bid on and won the job. This included maintenance, operations, and general support. It took years to train and develop someone with the unique skills needed to do their jobs. Finding someone willing to do the work under the conditions found at the range could be difficult.

An Up-range Threat System   (USAF Photo)

As a result, when it was time to bid a new contract, the companies who planned to bid would hold interviews in a Las Vegas trailer park in order to get names and contact information for these unique people. It would have been foolish for a company to assume they could start from scratch. Each contractor developed a list of skilled operators and maintenance techs. It was prearranged, and the company that won the contract just gave the crews a new company hat on the way to work the first morning.

One communications support company was Ford Aerospace, which no longer exists. The civilian contractors knew all the ins and outs of maintaining complicated communications equipment that controlled the threat array. They kept the range going. Site-to-site radios, range-to-Nellis telephones, threat-to-range control links, truck radios, and temporary nets; it was all theirs. I could always rely on them. I hate to say it, but so much of this kind of teamwork and cohesiveness seems to have been lost today.

A day in the life of ...

Driving around could be an experience. The threat sites were scattered around the range floor and it could take a while to get from one to another. In the morning the air would be pleasant, even cool; no one had yet disturbed the overnight quiet and stillness. Not all of the little residents had returned to their daytime shelters yet. Most interesting were the little sand-colored lizards who always stopped for a look at what I was doing. Were they as fascinated with me as I was with them?

At the threat sites, operators would be getting ready for another busy day; making sure the systems were ready, calmly aligning or frantically fixing their equipment. Range rats were particularly fixated on always being ready for a mission. There were few times a threat was not fully or at least partially ready for a range session, and almost never did a threat have to sit out a whole day. About the only time a particular system missed more than one day was for a scheduled upgrade.

Besides the equipment, there was all the operational stuff. Mission schedules, operator logs, videotapes; the area around an operator's console could look like the surgical suite of a hospital, with everything laid out just so. Most crews brought their lunch to the site rather than take time to drive back to the main area to eat. So in addition to operational supplies, most sites had a small refrigerator, a hot plate, and magazine collection that rivaled most dental offices for staleness.

At first I thought they would resent this mysterious captain poking around their site asking dumb questions and in general getting in the way. Not so! They had so much pride in what they did and had so much respect for what they produced that my time up range was a real joy. At the end of the day, the dust covering me head to toe felt like a badge of honor for associating with such a special group in such a special place.

Eyes of the threat (Photo: U.S. Air Force)

There was one side effect from the sun, which could be intense summer or winter. I grew up in the Deep South and spent my youth splashing in the waters along the Gulf Coast. This was before doctors started paying much attention to what this did to the skin; or I did not pay much attention to the doctors. Add in my years in the desert and my dermatologist has sent his kids to college treating lots of pre-cancers and occasional skin cancers; mostly on the left side of my face and my left arm, parts that were usually exposed to or stuck out of a truck window while wandering around the Nevada desert.

The Protectors

Because there were so many special sites and projects up range, the Air Force and Department of Energy in those days contracted with Wackenhut Security for protection and range access control. This was before contracting became a dirty word due to self-inflicted wounds by companies and general stupidity in government contracting offices. They also served the Department of Energy and others.

These guys and gals were real pros. I never had anything but positive, productive interactions with them. It could be fascinating watching them during practice security drills or listening to stories about some of the things they did.

One area I found especially interesting involved the range targets. Most were fitted out with old junk vehicles. Once in a while some thieves would sneak onto the range to scavenge parts from trucks and vans in storage areas. Supposedly a couple of times they hit targets, not real smart with the possibility of explosive laden duds lying around.

Especially popular were axles and wheels hard to find on the open market. The guards found stopping this bunch a point of pride and good practice, often doing it on their own time. They were very effective; and this was before night-vision goggles.

There were questions about how much of an official big deal to make of petty thieves stealing junk. Either way, apprehending or just scaring the hell out of them and chasing them away to spread the word to others with similar designs was usually enough. No one ever came back for seconds.

This was an indication of what would happen if anyone ever tried any serious incursion on the range.

Holstering my new guns

My first encounter with the threat simulators came even before going up range. During a tour of the Range Group maintenance yard at Nellis, the first thing I saw was several M-114s being renovated for up range. Newly modified F-5 radars and other electronics were being installed. I thought these were the coolest machines I'd ever seen, and was excited by the idea that these were some of what I'd be hunting with; the deadly Soviet ZSU-23/4 anti-aircraft gun with everything but the bullets, using a TV camera instead. These scout vehicles never caught on with the Army, but proved ideal for Red Flag.

ZSU-23/4 Shilka
(Photo: National Museum of the Great Patriotic War)

159

Up Range

Wandering through the workshops was impressive. It was amazing to see what they were working on. This was at the dawn of the computer age when most of that the engineers and technicians were doing was in the analog world. A big focus was on making it possible to simulate more than one type threat with a single unit to add variety to the missions. In those days you could not go to a threat simulator catalog to order "one from Column A and one from Column B" to put a range together. That would come years later.

Up range, the maintenance facility was straight out of the movies where mad scientists bring to life all sorts of creatures. Actually, they were doctoring circuit components and building others, as well as modifying various shelters and antennas so they would work and look authentic pilots' on-board warning systems. They then turned the stuff over to the operators who went the rest of the way and used experienced, nimble fingers on knobs and dials to perfect the signals radar warning receivers would see. The occasional use of a sophisticated signals collection van that could accurately measure signal frequencies and evaluate waveforms, determined how accurate the signals generated were.

So in addition to my attraction to the scenery, landscape, and atmosphere of the range I had great admiration for what went on up there. I quickly caught on to the fact that as long as I had these pros backing me up, I need not fear making bad shot calls for the fighter jocks and the electronic environment they faced would be accurate and realistic.

Pilots who had a chance to visit up range were always impressed with the operation. Depending on the setup, a couple of threats would be off line for part or all of a session. This was one way to tailor the EW picture attackers saw. It was how we accommodated an intelligence picture that said the enemy was adding defenses around a target by moving threats into that area.

At a couple of sites the crews put a lounge chair on top of the operator shelter. It was a great place to sit with a cuppa and watch the air show when their system was not in the lineup. All the attackers had to do was figure out how to see which threats had crews sitting on the roof to know what they would not face.

Low pass over a threat. (Photo: U.S. Air Force)

Sky Rockets in flight

Pilots praised the authenticity of the threat array and its impact on the mission – especially the Smokey SAM visual simulator. In the cockpit it combined an aural warning from the radar warning receiver with the sight of a smoke trail coming up from the ground. This added a major touch of reality to the training.

This was one of the most inspired and useful of the range creations. Cooked up at NAS (Naval Air Station) China Lake, California, it looked rather like a fat reject from the model rocket section of a hobby store. *Smokey SAM* was a stubby cardboard tube with a rounded Styrofoam nose and plastic tail fins. Filled with a very smoky propellant, it would be manually launched from a simple wire stand, either in conjunction with the initiation of a SAM launch signal or without an electronic tone from areas where intel briefs said shoulder-launched MANPADS, heat-seeking missiles, were supposed to be.

Pilots who had flown in Viet Nam reported that their most crucial defense maneuvers came as a result of both the audio warning from their protective systems and seeing the smoke trail of the missile. One pilot told me after his first encounter with a *Smokey SAM*, "my asshole has not puckered up like that since dodging SA-2s in 'Nam." Even today, they are considered one of the most valuable aspects of Red Flag range simulator operations.

162

As with just about everything Red Flag, it comes with a story. I was not involved in their invention, although I did help develop strategies for *Smokey SAM* use. To be successful they had to: one, work; two, be safe; and three, look right.

They were simple devices, so getting *Smokey SAM* to work was easy. They were designed to disintegrate without doing any damage if sucked into a jet engine. That addressed the safety part. They were a natural for looking right as long as the operator set them off at the proper time and they were located realistically close to the signal source.

One tale lived on from the visual tests during its development. Observers were watching launches from a UH-1 Huey. In one test the ground crew accidentally(?) shot one at the helicopter. The story was that it came in an open door. According to someone who was involved, "Luckily we had the other side open, so the damn thing zoomed straight through. We'd have been in a real mess had that other door been closed, Smokey came in, and the little devil (not the exact word they used), started buzzing around looking for a way out." Some may think this a little far-fetched; but after associating with this bunch up range, one has to think there may well be a lot of truth there.

Coming up with techniques for using them was pretty easy and the threat operators got the hang of it quickly. The rule was to shoot close-to, but not at, a plane. I'll forever deny that there was a case of beer for the crew with the closest shot during the first exercise that used them. I did apologize profusely to the pilot who had one bounce off his cockpit. He was so impressed that he tossed in a second case for the crew. Once word got out, there were several requests to specifically target select pilots, with squadron commanders being favorites. This was fun while it lasted; but it was strongly suggested that once was enough. So that was that.

One incident, however, could have had really serious consequences. One of the threat/*Smokey SAM* setups had the launch operator sitting in a beach chair with the launch switch and connected via headset to the operator inside a ZSU-simulator. I happened to be driving around the range on some task or another when over the range net came a lot of shrieking and screaming, not all of it understandable or polite. I had no idea what was going on, but it sounded like something had gone badly wrong.

Whatever it was, somebody was going to want to know the details, so I decided to start sliding in the general direction of whatever it was. Sure enough, the Red Flag Director of Operations came on the net and told me to head to a specific site and check out the report that it had been bombed.

When I got there, what I saw nearly have me a heart attack. There was a bent fender on the rear of the threat simulator, a bomb then hit the concrete pad the simulator was sitting on, gouging the concrete and slicing through the operator's headset cord. It then bounced off into the desert. Fortunately, it was a 'blivit', a 500-pound bomb filled with concrete rather than explosives.

The DO then radioed me to go find the bomb and report the serial number so they could determine who the hell dropped it, and do it fast. I took off on a dead run through the mesquite, sidewinder rattlers scrambling out of my way, wanting nothing to do with this crazy GI stomping through their area.

Sure enough, several hundred yards into the desert I found it, buried into the sand with a bent tail fin marked with the same color paint as the threat fender. I got on the radio to Red Flag to report the results of my search. The DO told me they no longer needed the numbers because they found the dummy (loose translation) that bombed a threat site instead of the assigned target.

Over the radio I could hear the Red Flag Commander in the background in a very loud uni-directional roar describing a particular pilot's intelligence (lack of), skill as a fighter pilot (even less of), and general unsuitability to even be on the planet. I wondered if by the time I got back to Nellis the next day there would be a vacant room in the Visiting Officers' Quarters. Red Flag did not tolerate such stupidity very well.

The Sandbox

The layout of the threat range was interesting and clever. Spread across an expanse of the desert floor, close enough to the targets to be realistic, far enough away to reduce what could be fatal mistakes. A pilot was bound to cross swords with at least some threats, no matter what the target. They were also situated so there was opportunity to practice terrain masking techniques during their attacks as well.

Since I was seldom out on the range during a full-up Red Flag mission, I developed a deep appreciation of the smells, feel, quiet, or roar of the place. Driving from one site to another could take up to half an hour, a great way to enjoy the solitude if things were quiet. Better yet was taking care of business up at the mountain top communications sites far removed from the busy range floor with the trees, birds, and wildlife for company.

His view of me? (Photo: U.S. Air Force)

That is not to say it was always quiet. When Red Flag was between exercises, others used the range. There was a 300 foot low-altitude limit for operations. Little did I realize 300 feet was so close to the ground.

One day I was driving in the center of the threat area when straight ahead, from where the road met the horizon was a dark spot that kept getting bigger and bigger and bigger and BIGGER. I slammed on the brakes and bailed out of the truck as a B-52 passed right over me. The pilot had to climb so he could dip a wing and turn. I swear that besides the jet exhaust I could smell and feel the heat from the plane's skin.

There were other times when some helicopters hop-scotched around in the distance. I didn't know what they were doing; but given some of the inhabitants of parts of the range, it was better that way.

The Wild Weasels did some serious jousting with the threats when they had the range to themselves. It was a treat to see them work their targets, especially in pairs. But my scariest experience up there came one day when I stopped and walked up a small rise to get a closer look. It was a somewhat hazy day and a pair of F-4Gs was really giving it hell.

They were going at one site big time. I could hear my boots crunch the sand as I climbed up to the top of a hillock to get a better view. Just as I topped the rise, I was suddenly nose-to-nose with a Weasel blasting along in what could only be described as a serious manner. I flattened myself face-first on the ground and felt the wind from the plane blow my cap off. Damn near wet my pants on that one. He climbed up, swung around, and rocked his wings as he passed – either saying "sorry" or "gotcha".

Crazy as it may sound, I can't be around low-flying planes without getting some of the intense feelings of that day back. Damn that was fun!

Weasel at work (Photo: U.S. Air Force)

Playing with the Big Boys

B-52 bombers were a serious challenge to the threats with their powerful and versatile jammers. Countering them was nearly impossible; but the bomber crews appreciated the chance for some serious training against real transmitting threats. The only thing a B-52 carried more of than bombs seemed to be jammers.

The crews did wish they could land and debrief with the operators face-to-face regularly. That would have been great. The threat operators hated never being able to get through the jamming; but admired the power and skill of the bombers and their EW Officers.

B-52s were not always part of a Red Flag. One day one landed at Nellis for a visit and we had an opportunity to see it up close and talk with the crew. I enjoyed crawling around the cockpit and EW compartment.

I noticed how wrinkled the skin was looking down the side of the airplane. The crew explained that pressurizing the plane at altitude straightened the wrinkles right out. During a recent visit to Seattle, Boeing had a piece in the local paper about planned modifications that could extend the life of some B-52s to up to 70 years.

B-52 over the range. (Photo: U.S. Air Force)

Sneaky, Sneaky

During one Red Flag I hatched a plan to shake things up a bit and poke a hole in aircrew complacency. An airman operator and I took an SA-7 simulator and hid in the rocks where the planes crossed into the range (Coyote). This followed days where the intel brief included warnings that hostile forces were operating in that area, moving forward to the edge of the front line, and were equipped with MANPADS.

This turned into quite an adventure. The best spot turned out to be a set of ridges just outside the range boundary where attackers turned in for range entry. They had just passed through the Red Flag CAP

(Combat Air Patrol) area and had to switch from air-to-air to ground-fight mode. I must admit, the two of us also took up combat-ready mode, even when we were the only two warriors sneaking around the rocks. It must have been quite a sight, us hunkered down behind some boulders with our MPADS simulator and semi-battle-rattle outfit of back packs filled with video tapes.

We engaged everyone successfully, threw the tapes into a truck, and made a mad dash to Nellis for the mission debrief. The first time just about everyone saw themselves in the cross hairs wandering nonchalantly into the fight area. The pilots were not happy, drawing an "I told you so" from me. I emphasized that the presence of the threats were in the intel briefs they did not seem to have paid attention to.

The SA-7 Strela (Photo: militaryfactory.com)

After two days starring on SAM TV, they caught on and hit the range yanking and cranking as they crossed the line. One F-4 was working so hard it nearly smacked into a ridge.

Bad Times

Combat, here's a surprise, is dangerous; even with simulated bullets and missiles. Red Flag is no exception. There were some bad, fatal, and nearly fatal incidents during my time at Nellis. During lunch in Harry's Bar, on the first floor of the aged Hotel Harrington just blocks from the White House and around the corner from the FBI (so you can imagine what the lunch-time crowd is like) I discovered a painting of a B-52 nearing Jackson Peak and entering the range – game on. It jarred my memory back to a major crash experience during my time at Nellis.

It was mid-morning and a B-52 was scheduled through the range on a routine training mission. Everyone was ready for what was to be a simple in-through-gone event. They happened all the time for bomber crew training. The bomber's ingress time came and went; no contact. Five minutes, ten minutes, twenty minutes – nothing. Range control kept getting more and more tense, burning up the phone lines back to the bomber's home base and saturating the airwaves with radio calls. The DO began to call and organize a search mission to find out what happened, hoping against hope that there was a simple explanation.

Then came word from the Nevada Highway Patrol that there had been some sort of major crash on Jackson Peak. B-52s used this entry point, terrain masking behind the top ridge, popping over and onto the range where they and the threats spent ten minutes spraying one another with electrons until the big bombers exited the range on the other side and swung around for home. A search team verified what we feared; the big bomber slammed headlong into Jackson Peak just feet below the summit. The impact must have been horrific. Investigators on the ground reported that nothing larger than a Volkswagen would be found scattered around the impact point. Blessedly, what happened to the crew would have been fast.

One unique feature of crashes at Red Flag is that if they happened on the range, there was a good chance that the incident was caught on video tape by one or more simulators. My first was when a Royal Air Force Buccaneer unexpectedly rolled into the ground when a wing collapsed during a target attack. Rescue forces were immediately dispatched; but nothing could be done. After the crash I worked with the British squadron commander and investigations officer to get things sorted out. The range threat video would be important to investigators' analysis of what happened. So would interviews with the operators who observed what happened. In this case, I also got involved with the RAF Flight Surgeon who coincidentally had been hospitalized with appendicitis. It was up to me to ferry the paperwork back and forth to his room; a fascinating experience, at all hours day and night. It was interesting to see what sort of information was gathered on the bodies of the victims. That can be even more detailed than the investigation of the airplane.

Another RAF crash came a little over a year later when a Jaguar was caught on video from one of the AAA simulators, so there was a good record of that event. The threat operator was very professional and did a great job tracking the aircraft through its approach and apparent control malfunction while attacking a target, resulting in a nose-first rolling crash behind a small hillock. The up-range crew rushed tape and crew logs back to Nellis with a little help from a helicopter. On instinct, I made an immediate copy for the investigation, preserving and guarding the original as evidence for the Board of Inquiry. In those days, every time a section of tape was viewed, a little quality and detail was lost.

The whole thing proved quite upsetting to the operator who was tracking the Jaguar at the time. She became very concerned about the pilots' families, including starting a collection of plush toys for his children. The British team ended up giving more comfort and sympathy to the operator than vice versa. They also officially recognized the operators for their quick thinking which captured valuable evidence that directly contributed to the post-crash investigation. It turned out something snapped in the plane's controls causing a left roll into a nose-down crash into the ground.

It was interesting to watch range control in action during these emergencies. There was a basic plan for such situations, but the range was so big and operations could be so complex that lot had to be made up on the spot. When these kinds of crashes occurred, there were a lot of airplanes in the air around the site. Once the "knock-it-off" goes out over Guard (the military emergency radio channel – 243.0 MHz) the exercise stops and everyone wants to help. But with so many aircraft in the area, letting them mill about trying to help could result in mid-air collisions.

It was impressive to watch the weapons controllers at their radar scopes immediately access the situation, evaluate what resources were where, and sort them out so some aircraft covered the site to see if there are survivors and where they were so rescue forces could be directed to them. Others were assigned top cover to keep the area over the site clear of interfering aircraft, and the rest were told to come home. Other controllers came on-line with rescuers to coordinate their flight to the scene, clearing the way as needed en-route and at the site. The Senior Controller kept quiet, letting the weapons controllers at the scopes do their work and making sure they got any support they needed. My job turned out to be polling the threats to find out if anyone had witnessed the event and letting investigators know what eye witness information was available.

Fortunately, not all accidents resulted in the loss of a crew. Once, an F-4 was engaged by a threat and the protective system launched its flares. Unfortunately, the outer doors of the flare dispenser did not open; so the crew found themselves flying around with a batch of pyrotechnics burning <u>inside</u> the rear of their plane. This was not good. So they pointed the fighter in a relatively safe direction toward open space and punched out. Once the pilot and back-seater got to the ground the exercise could pick up where it left off. Someone would mop up the pieces later.

* * * *

6. Maple Flag

It became obvious to planners that airspace restrictions around Nellis interfered with Red Flag operations. In spite of the amount of military-only airspace over the range, Las Vegas air traffic, cross-country airways, and local flights were an issue around the edges of the range. Mike Renyo, in his 1992 book *Maple Flag*, penned an excellent history of the birth of the Red Flag-like exercise in Canada. In 1977, the commanders of the USAF Tactical Air Command and Canadian Air Command started a bi-annual operation in the open spaces of Canada.

Nellis had the Red Flag staff, Canada had a vast area of free airspace around CFB (Canadian Forces Base) Cold Lake on the Cold Lake Air Weapons Range (CLAWR) in northern Alberta. With 3,000 square miles of area where ordnance could be dropped and 7,000 square miles of exercise space, Maple Flag could operate in an area the size of Kuwait. Multiple, variable ingress/egress routes were possible. Air-to-air combat could be conducted with minimal restrictions, often limited only by squadron rules. VFR conditions made a faster paced operation possible since fliers did not have to contend with crowded civilian traffic lanes. Not having to operate under IFR conditions made it possible to conduct large-scale, comm-silent missions. And the base had a lot of room to host visitors and their airplanes.

There were seventy tactical target complexes configured as airfields, truck parks, tanks, bridges, convoys, and AAA/SAM sites. The terrain was like Central Europe, with rolling hills, thick forest, and lakes. Locating targets in small clearings made acquisition and attacking them a challenge and valuable training for pilots. Overall, it

173

was a far cry from the constrained environment of Nellis and its range.

The first Maple Flag exercise was held in 1978. The Red Flag staff relocated to Cold Lake for the duration. Six weeks allowed for three two-week rotations, increasing the number of aircrew that received the training. By 1981, the Canadians had gained enough experience to run the exercise themselves, with Red Flag staff in support and providing specialty capabilities, like threat simulators that added live EW to the operation. In 1987, the frequency went to annual so the CFB-based squadrons had more time to concentrate on their own training requirements.

Maple Flag was not designed to see who can beat who among the participants. Instead they were given the opportunity to learn the tactics of an opposing force in a real scenario. It provided an opportunity to experience tactics and operations in large strike packages of over 75 aircraft, something younger pilots had never experienced. It was a valuable learning experience for Canadian, British, and U.S. aircrew in an environment where no one force dominated planning. With each group learning how the other operated, they would be able to operate together more effectively in the future if they had to. Other nationalities also participated from time to time at Maple Flag. Some air forces took advantage of being on the scene because they were at CFB Cold Lake for routine training or testing and talked their way into being allowed to participate from time to time. They were always up for the parties.

Like Nellis AFB, CFB Cold Lake was born because of a need for a place to develop and test weapons away from populated areas. In 1951, Canada decided that an area 33 miles deep and and 104 miles long on the Alberta-Saskatchewan border was ideal and opened the Air Armament Evaluation Detachment (AAED). Over time, CFB Cold Lake grew into Canada's "Fighter Town". Every fighter ever flow by Canada has been stationed at Cold Lake at one time or another. It was considered one of the most modern fighter bases in NATO.

CFB Cold Lake ramp   *(Photo: Royal Canadian Air Force)*

Way Up North

It did not take long before I really got to like Maple Flag. Packing up and flying our mobile threat simulators, with all the associated radio, TV, and communications gear, was hard work. But the end result was worth it. In addition to the actual exercises, there was a short planning conference proceeding each one. That doubled my Canada time.

The threat group and survival instructors ended up forming a small posse that got hooked in with some of the locals; becoming a family that looked forward to our regular reunions in Grand Center, the local town. The home of the head the local Royal Canadian Mounted Police detachment became our field headquarters and was transformed into the "U.S. Consulate" by displaying a special brass plaque in the front window. It became a staff party HQ, and we were declared "half-ass Canadians" by our friends up there.

Maple Flag

In Grand Center, I became great friends with a family that owned a new motel built in time for my second Maple Flag. Many of the staff were assigned rooms there. The owner's wife figured out that if her husband was gone too long he could probably be found in my room because "something needed to be fixed." This never occurred until I'd had a chance to stock the 'fridge with LaBatts Blue. Naturally, hospitality dictated I offer refreshment when he came to help. The last night of my last Maple Flag found me in their apartment with them, the kids, pizza, and beer. Since this time I would not be coming back there were tears all around before the night was over.

The Canadians were so kind and hospitable that once when we arrived in Grand Center in the wee hours after major delays in driving up from Edmonton. They opened the cafe and bar at the hotel where many staff would be staying. In spite of the hour, food and drink aplenty was available. The conversation got around to fishing (that never took long in Canada). Based on a off-hand remark, one of the guys started calling his buddies to come take us fishing. At 3am!

An on-going excitement for we Americans was the chance to see the Northern Lights during the winter planning conferences. We would always beg our Canadian friends to take us to the shore of Cold Lake to see them. We were thrilled at what we seldom saw; our friends got cold watching what they saw every night. Canadian hospitality being what it was they never turned us down. But after a while they got smart and we could go only if we brought a nice stock of whiskey along "to fight the cold."

Downtown Grand Center had an actual curling rink. I was introduced to the intricacies of pushing a really big, really heavy rock along a strip of ice and frantically sweeping the ice ahead of it. I'm sure there was a point to it all. Once I got the hang of sweeping like a fool without slipping on my butt it got to be fun.

Once, the town decided to throw a big dinner for everyone at Maple Flag. The only place big enough was the local arena. They drained the hockey rink so there would be a place to eat.

We always looked forward to the drive from Edmonton to Grand Center through the stunning Canadian wilderness. It was a time to appreciate the northern woods, deep green trees on both sides of the road, creating a tunnel of relaxation.

Part of the Red Flag staff, including me and the threat crews, always came to Edmonton early to set up the rental cars for the exercise. It took a lot of autos for all the crews, so this was a business boom for the agency. We would pick up all the keys and go back to our hotel to sort them for the different squadrons and other users who would have someone pick up their car and drive it up to Cold Lake. The rental agency owner appreciated the business and usually made sure we had adequate refreshments in the trunk of our car for attending to the key sorting task. I'm not sure how legal that may have been; but midway through sorting dozens of keys it surely was appreciated.

One time, the owner stopped me and my crew on the way out of the gate, saying "there was a problem and he had to change our car". He came out with a Lincoln Continental to replace our Chevy sedan that "broke." Fine with us. About a week later up at the base the Red Flag Commander said that he'd gotten tired of us having a limo while he was stuck with a Canadian Forces pickup truck (because it had a flight line radio). He made us turn our Lincoln in for a regular car. He was really a great guy and did that just to mess with his threat team.

Later over drinks at the Mess, we had a good laugh as the colonel described the look on our faces when he nailed us. Oh well, it was nice while it lasted. I have to admit, wrestling the Class-A Land Yacht instead of a regular car around the area was getting tiresome. Besides, how could we enjoy flinging stuff into the luscious leather back seat of the Lincoln with the same abandon as we did in the Chevy?

Life was always interesting

Traveling into Canada regularly always seemed to involve some sort of 'interesting' event. As regulars, we got to know the Canadian Customs and Integration officials and they got to know us. We traveled with Diplomatic passports to speed up and simplify our passage through the gateways. For some reason, to the great delight of the rest of my crew, just about every time coming through Edmonton, airport security pulled me aside and one of the female agents at the screening point had to wand my nether regions. She always got a big laugh out of it too. It could not possibly have been a setup by my crew, could it?

One trip up on a USAF C-141 brought us straight into CFB Cold Lake. It was the last Maple Flag for the Senior Master Sergeant Rescue Detachment supervisor. He had been with the exercise longer than anyone else and a great friend, especially to the Mounties of Grand Center. He would also have a birthday while there. That cried out for something special.

On board the C-141 was a contingent of maintenance personnel. They and the subject NCO were about the only ones not in on what was to happen next. While we were taxiing toward Base Ops an RCMP patrol car, red lights flashing and siren wailing, signaled the pilot to pull to a stop. The C-141 stopped and the Mounties stormed aboard, handcuffing the subject and dragging him away. The maintenance troops were absolutely aghast as the police sped away with the NCO. We were not able to hold the laughing very long.

Our rescue specialist finally made it back to staff quarters the next day, quite the worse for wear from one hell of an advance farewell party. Since he did not have any real duties until later, there was nothing lost and US/Canadian relations got a good, if very local, boost.

Not everything of interest took place on CFB Cold Lake. One exercise included an HH-53 for the SAREX missions. It had to come from home base, hopscotching across Canada. The day and time of their arrival came and went, no helicopter. Time dragged on; no word. But since there were no reports of a crash or any disaster, everyone assumed there was some sort of breakdown or bad weather along the way and the crew was not able to contact Cold Lake. Finally, well into the next day, over the horizon comes this HH-53 looking far from on top of things for some reason. The helicopter seemed to limp to the parking apron and power down. Everything seemed to sag.

SAREX Jolly Green (Photo: U.S. Marine Corps)

The crew opened the hatches and it immediately became obvious something had happened. That was one sorry-looking bunch. The DO, after a some prodding, got to the heart of the story. As I said before, Canadian hospitality can be second to none. It seems that the last stop was at the Canadian Forces base at Moosejaw, home office for all Canadian Forces' helicopters.

The story, as best it could be pieced together was as follows:

1. After landing, the pilot sent the pararescueman to the other side of the facility to get some maps for the exercise. This entailed passing through several offices filled with Canadian helicopter crewmen.

2. After an hour or so he did not come back. So the Crew Chief was launched to find out what was taking so long.

3. Another couple of hours, and now two crew members were missing.

4. Time for the co-pilot to go investigate. He never came back.

5. It was now time for the pilot to find out what sort of black hole was sucking his crew into oblivion.

This is what the sorry-looking crew explained. The pararescueman had to pass through the offices to get to the map room. At each office there was a toast for luck when they got to Maple Flag. Repeat for the crew chief and co-pilot. According to the pilot, by the time he got to the map room, there was one roaring party underway. That put an end to the rest of the trip that day.

Heads pounding, the Jolly Green crew made it to Cold Lake. Even the helicopter looked hung over. The crew hoped to crawl to their quarters for some badly-needed sleep. The Commander decided instead that this would be a good time for a series of inspections and checks on the helicopter, a process that would take the rest of the day. When it came time to quit for the day, the crew politely declined several invitations to after-work drinks at the Officers' Mess. I wonder why.

Our Northern Office

Threat control set up shop in an old, abandoned control tower in a corner of the parking apron. It was a great place from which to observe field operations and gave us ample room to work. One exercise, the British AV-8B Harriers parked next to the tower. For some reason, there were frequent problems causing a fuel spill, necessitating a response by the base emergency crews. We got to be pretty friendly with the firefighters, who, once they finished taking care of the mess, dropped by to see what we were up to.

This was one of the last places in the world to see operational F-104s. They had a batch at the Evaluation Unit.

Seeing tanks fly

One major effort before the exercise could start was getting our threat equipment up range. This involved a variety of stuff, much of it in large metal cargo bins. Most interesting, though, was the mobile gun simulators. These had to be slung under a Canadian CH-47

Chinook helicopter because there was no way to get them inside. "The day of the flying tanks" was always popular with the locals, especially the school kids. I would let some Grand Center teachers know when this would happen so they could let their students out on recess as we passed over town and the kids could get a big kick out of the sight.

These Guys Make It Look Easy   (DoD Photo)

It became my task to climb on top of a threat and hook a sling to the helicopter using a hook extended through a hole in the bottom of the Chinook. That way, if anything went wrong and we dropped a simulator, I guess I and not the Canadians, would have to pay for it.

A key part of this operation was to plan for an emergency during the hookup and liftoff. After a lot of coordination with the pilots and load-master, we decided which way I would dash to safety. The pilot would pull the helicopter the other way. That is also the direction I would go once hookup was complete so I would not be under the load when the helicopter took off. Inevitably, just as I got out from directly

181

under the CH-47, the pilot would pulse the cyclic to create a burst of down-wash that would blow me ass-over-teakettle into the woods next to the field. They got their laughs, but I usually got a couple of free beers that night at the Mess.

Cold Lake Threat Site   (Photo by the author)

The threat operators were masterful in the way they could create an authentic-looking environment for the attackers. Nimble fingers changed frequencies and waveforms to make a couple pieces of equipment appear to be a wide-spread threat array.

Unlike Nellis, there was no master range control net, so the operators functioned relatively independently. They had an IFF/SIF transponder to give them the position and help them track incoming aircraft so they could engage effectively. Not realistic from the impersonation aspect, but necessary for the training side.

One time an Air National Guard Tactical Control Group deployed with an AN/TPS-43 tactical radar. Good training for them and a help in creating more realistic threat network scenario.

Maple Flag Threat Crew   (Photo by the author)

At Maple Flag I had more flexibility than at Nellis. While the Nellis range was relatively fixed geographically, in Canada the size of the Cold Lake range made for a wider variety of mission geography. Targets could change, ingress and egress routes would vary, and how we played on any particular day was flexible. There were some days the threats were not a part of the entire exercise that morning or afternoon; but usually there was some part of the package assigned a task that brought them into contact with the simulators.

My time was not as jammed as in the States. Getting threat results from the simulator operators was much more timely. We got a lot of assistance from the Cold Lake rescue detachment. Usually they had a

UH-1 available to transport people and our threat results to and from the site quickly, so I could often have the results of a mission in my hands and ready for analysis as the last of the aircraft were landing.

The simplicity of the threat operation was a help from that standpoint. Since there were far fewer simulators than at Nellis and their tasking was uncomplicated, the operators would send me results that were just about briefing ready. All I had to do was to prepare the scoring slides and pick which engagements made TV that afternoon.

The Author in Star Wars Control Mode   (Photo by the author)

As a result, depending on how things were organized on any particular day, I sometimes had time on my hands. Some of that time would be used talking to crews about threat tactics. The rest of the time could be used doing other, fun things. I flew with the AWACS crews supporting the Blue Force, I once rode as spotter in a Canadian C-130 during one of the biggest peacetime air battles ever, all done in radio silence. That was impressive.

After the C-130 ride, there was a scene that will always stick in my mind. That exercise scenario created an enormous, swirling fur ball of aircraft yanking and banking like there was no tomorrow. Had it been real, that would be true for many involved. Anyhow, before it was over just about everyone, even the most experienced fliers in the C-130s, were airsick.

I flew in the navigators seat as spotter and photographer. The donnybrook would have put the Battle of Britain to shame. I got airsick, a rarity for me, when I looked down to change film in my camera. I did not get any decent pictures and spent the rest of the flight various shades of green.

One of the C-130s carried a Cold Lake air traffic controller who had been rewarded with a chance to fly on a mission. We landed and made it to the parking area where three quarters of the flight crew jumped out of the doors and headed to the edge of the ramp. I'll never forget the sight of all these experienced aircrew throwing up in the grass while the controller went happily skipping across the ramp to the Ops Office, completely unaffected by the airborne gyrations, her blond pony tail swinging in time with her steps.

One of my range rides   (Photo: Royal Canadian Air Force)

Maple Flag

In general, I spent a lot of time poking my nose into things I never had a chance to at Nellis. Interestingly, I was welcome just about everywhere in Maple Flag and occasionally a few places that had nothing to do with the exercise. The only rule was that I cause as little trouble as possible, and most of the time I was able to abide by that. But given the Canadians' spirit of adventure and enthusiasm for mischief, I suspect there were times when they took advantage of my presence so they could have fun, and blame it on "the American who did not know better."

Unlike Red Flag, I had the advantage of being able to go up range to observe the operation because a helicopter could transport me to and from the threats much more quickly that I could get to and from the Nellis ranges. This helped me add color and context to my briefings.

Once, the helo dropped me off. This came after a couple of days scoring many 'kills' against the A-10 squadron participating in that exercise. They appeared on a lot of my debrief videos.

That day I was standing atop the hillock behind the threat. 'General Patton'-like at one point, I was standing there waving a big stick (a branch without the swagger). The A-10 pilots got even for being selected for the video threat debriefs by what I have to admit as a really clever strategy. One was flying directly at the threat, low and weaving side-to-side effectively. I could tell from the way the antenna was flailing that the operators were having a hard time keeping him in their cross-hairs. Suddenly I heard a loud buzz behind me and spun around. There I was, looking directly at a GAU-8 that popped over a ridge just behind the site.

I turned out to be the star of the pm debrief that day when the squadron did their presentation. The deer-in-the-headlights look on my face greatly amused all the pilots I had been showing on the threat video every day. It cost me several cans of Lebatt Blue that evening. A couple of very clever A-10 drivers had earned them.

After one mission, an A-7 returned after flying too low on a strafing run. There was a log embedded in the leading edge of the right wing. Staring up at the plane, the safety officer said, "First time I ever saw a flying Class A (totally destroyed aircraft)."

At Maple Flag, SAREX participants were also selected based on threat results. The underlying idea was that a pilot was more likely to be able to survive getting shot by a threat while flying close to the ground than by an air-to-air shot at altitude. I was friends with the survival instructors and base rescue people. So in Maple Flag logic it only made sense that I should ride along with the chief instructor who went up to evaluate the 'rescue' effort. There was some socially redeeming value to this. I got to watch flight operations up close; plus for this ground-pounder it was fun.

I did get involved in what may be the only UH-1 vs. F-15 dog fight in history. We were flying through the valleys when a Combat Air Patrol F-15 spotted us and came down to investigate. Naturally, the helicopter pilot was not going to let this opportunity pass by and issued a challenge to the fighter jock. So for the next several minutes we dodged around in the valleys turning nose on to the fighter every time he attacked, spoiling his shot.

Finally, we ran out of places to hide and that night's gun camera video at the debrief showing what happens when a mouse roars in a dog fight. We were given lots of points for guts, not many for smarts.

Things go real wrong real quick

While most of these adventures were fun, there was one time things went real wrong real quick. It was a beautiful day and things had gone well with the survivor pickup. Top cover setup was right, the coordination with the search and rescue force went well, and so did the 'snatch and grab'. On that flight it was the helicopter crew, myself, the SAR instructor, and an add-on, a flight examiner for one of the helicopter pilots who would log a check ride.

We were happily motoring back to base when over the radio came the dreaded "knock-it-off, knock-it-of, knock-it-off. Airplane down." In unison it was "oh shit!" That call never meant something good had happened. A quick check and we saw smoke from the trees a couple of miles behind us.

Cold Lake rescue unit helo   ***(Photo: Royal Canadian Air Force)***

An F-4 on a strafing run did not pull up in time and flew into the ground right at the target. We radioed and volunteered to help, since we could do a 180 and be at the site in minutes. There was a quick switch from the SAREX call sign to a real-life rescue call sign and what started the day as a practice search and rescue suddenly became a real one.

The Huey was on scene quickly and we could tell it was bad. Flying down a swatch of trees broken off increasingly closer to the ground and a little wider that an airplane's wingspan it became obvious this was the last path of the fighter. That was settled when we saw a major pieces of two wings crumpled in the trees on either side. Smoke was everywhere and getting thicker the further in we got;

white, gray, and black with angry red patches of fire deep inside the cloud.

The pilot landed close to where the wreckage began. It would have to be on foot from there. Things looked grim, but we held out hope. What if the crew ejected at the last second and were laying in the woods around the site injured? We had to try and find the pilot and back-seater so we could bring them to safety. A quick conference and it was decided that three of us, the survival instructor, myself, and the flight examiner would be the ground party. The helicopter would fly to a nearby fuel cache to refill the bird's fuel tanks and come back. We grabbed what first aid supplies we could and hit the ground.

As the helicopter faded into the distance and the noise of the rotors died down, the three of us were engulfed in a world of intense sights, sounds, and smells. The smoke stung our noses, the trees crackled as they burned, and heat began to attack our faces. The stench of burning wood and jet fuel was intense and ashes floated down on us.

A Bad Time Indeed.   *Source: U.S. Air Force*

As we made our way closer to what was obviously the heart of the wreckage a different sound stood out from all the rest. Pops, zings, and pings. Once a couple of what sounded like buzzing bees zipped past our ears and we realized that ammunition from the plane was cooking off from in the heart of the fire and flying out of the wreckage, a complication we really did not need in what was already a scene from hell.

As the nightmare unfolded, we reached the first major pile of wreckage. There was the first indication that the crew's survival was not likely. It was where we found the first victim. Because crews fully simulated combat, they wore no name tag or other identifying markers. Instead of being too explicit about the grisly scene, let it be said that I was able to remove a wallet from one top pocket of the man's flight suit. The wallet would be the only chance to identify the victim, an all-important part of any crash investigation. I opened it to make sure there was an ID card and looking back at me was a photograph of his wife and two little girls. I probably would have gotten sick, but there was no time.

A trail of evidence spread along a path from where we were to deeper into the crash and an increasingly hot inferno. There were dollar bills strewn in a line and the three of us got increasingly concerned that there was someone in there. The wreckage was completely engulfed by fire, the shells were cooking off, and the woods around us were ablaze. One hell of a place for anyone to be, lying hurt after bailing out just before the plane crashed. Not a realistic evaluation of the situation. But given that we were more and more effected by the smoke and fumes, hope and adrenalin, rather than the evidence around us, that controlled our thinking. If there was anyone to be saved, we wanted to do it. But the blast furnace we were looking into cast real doubt on the likelihood of being able to do anything except adding to the toll.

The helicopter returned to where it dropped us off. "It's time to go – NOW!" was the message. The woods were in flames all around the scene and spreading into our escape route. It was not just the pilot and copilot's opinion, the word had come from the base to get back.

It was hopeless at the scene, and smoke was building up from the rapidly spreading forest fire, and would probably soon be interfering with flying back to Cold Lake. Further efforts would have to wait until a rescue team could be dispatched. Common sense helped us realize that this was the smart thing to do. So we stumbled, hacking and coughing from the smoke, our skin stinging from the blast furnace heat, back to the helicopter.

The pilot took off and made a mad dash for base. Smoke was drifting over Cold Lake and he did not want to get caught having to land in bad conditions. Besides, he said the three of us looked like hell and needed to be checked out as soon as possible. We were not feeling tip-top. It was all we could do to keep from passing out on the way back to base.

As soon as we set down, medics grabbed the Canadian flight examiner and our survival instructor and headed for the base clinic. The Red Flag and Canadian Base Commanders met me and we headed to the Commander's office to debrief. First priority was to report on the status of the crew. The wallet was crucial to verifying the status of what turned out to be the F-4's pilot. In any fatal incident, the Air Force puts top priority on making sure that any information released is accurate. No guessing.

Based on the evidence I could report from the scene it seemed most likely that neither of the crew survived. To insure no rumors got back to bases in the States to worry families with incorrect information or false rumors, outside phone lines were blocked for random calls. After providing information that would help rescue crews and the Air Force investigation team that would be coming I was taken to the infirmary.

This was before picture-taking cell phones. I was carrying a one-shot Instamatic pocket camera and had instinctively taken pictures of the scene when we arrived. Especially helpful would be the shots of the path through the trees, giving information on the angle at which the plane descended through the trees. I also captured what some of the wreckage looked like at the time. Much of this detail would be lost because much of the tree line around the site had burned by the next morning. Naturally, the camera became evidence for the investigators, and I never got back the other photos in the camera, including pictures of an irate beaver a few days earlier objecting to a couple of us fishing along the river in what he seemed to consider his private territory

I cannot say enough about the Canadians and the way I was treated. At Base Ops, the commander and his staff had the utmost concern for my condition. I will always remember the Base Commander's assistant hustling off and returning with some of the most delicious cool water to help my parched throat.

When I got to the medical facility, the other two guys had been treated. Oxygen, a good checkup, and a prescription for rest was called for. The medics said they would be OK in a few days. I got the same basic once-over. It felt good to sit back on a gurney sucking oxygen and unwinding a little. Cooling breaths from the oxygen mask felt wonderful in my lungs, although for the next several days breathing would be effected by my time in the smoke. Ever since, it seems from time to time I would feel the effects and have never been able to breathe as deeply as before; not a serious problem, just a lingering result. Running has not been an enjoyable exercise since.

A final incident that has etched itself on my mind came as I was being released. I got down from the gurney and brushed my hand across my forehead. There must have been ashes in my eyebrows because I felt something get into my eye. I said "ouch". Whereupon, with a shout of concern, the Canadian nurse slammed me back down on the gurney and literally jumped up and knelt on my chest, one of those lights doctors use in one hand and a syringe in the other. I remember vividly a very bright blue light, like those glow-in-the-dark Halloween costumes, and water being squirted in and around my eyes. She sheepishly apologized for the rough treatment, but explained that it was to prevent a cinder from scratching my cornea. Bless her. No apology necessary.

Everyone on the staff was concerned about us. After a much-needed shower, the Red Flag Commander took us to the Mess where he prescribed ample free drinks for the instructor and me to take our minds off the afternoon we had spent. It also would help me try and sleep.

During the night, it felt good to get outside and breathe some fresh, crisp night air from time to time. What was a surprise was the number of people who came out to see how I was doing. I guess they were keeping an eye on me. That's what the Red Flag team was like.

By the next day, search crews had found the other victim. He'd been killed instantly in the crash, so our fears of him getting caught in the fire were unfounded. They were able to recover both sets of remains. In a strange quirk of fate, the survival instructor was to re-enlist that day, so plans were made to have the commander swear him in in front of the base rescue helicopter.

Memorial To A Crashed F-4 And Its Crew.   Source: U.S. Air Force

The Canadians gave their flight examiner a commendation for his role in the rescue attempt. I was also able to get our survival instructor a medal for heroism. Both well deserved.

The forest fires raged on for days, and smoky conditions hampered operations for the rest of that Maple Flag. Many missions had to be altered because flights could not be planned into some areas and ingress/egress routes needed to be changed. They had to avoid investigators working the scene. The rest of that Maple Flag went well, all things considered. But it was always in the back of everyone's mind the dangerous business we were in. It was also a first-hand lesson on target fixation, which was found to be the cause of the crash. The mission was to gun a simulated tank. The pilot was apparently right on line because the plane hit the target dead center.

* * *

7. NDMCS

After I returned from the Telecommunications Staff Officer Course at Keesler AFB, Mississippi, I was assigned as Communications Project Engineer for the 554th Range Group. A range as large and complex as Nellis required communications; lots of it. A range control net, intra-range telephones, data lines, walkie-talkies, links to radios for aircraft using the range; it was quite an extensive setup. Then there were the security and classified links to various places that officially "did not exist". A good indication of how my day was going to go depended on whether or not by 0730 I got a blistering phone call from the north ranges or Nevada Test Site about system troubles. Plus, we needed to be able to add more threat sites and other data sources around the range.

All of this initially came under my jurisdiction. Fortunately, Ford Aerospace had won the contract to handle day-to-day operations, maintenance, and management. The techs were a group of capable, dedicated civilians who knew the range like the back of their hands. There was little they could not do, even if it seemed contrary to the laws of physics. Invariably, when I called the supervisor about something that needed to be done he almost always already had it on the schedule. The crews could find a quick solution to nearly anything that went wrong or unusual requirements that popped up.

Long in the Tooth

The old analog telephone system linking Nellis to the range was rapidly becoming unable to meet our communications needs. Adding more threats or supporting a hoped-for expansion of the mission debriefing system would need a larger backbone, new links around the range, and better reliability and stability. A plan was in the works, though; it was the Nellis Digital Microwave System (NDMCS).

Give Me the Money

The Tactical Air Command (TAC) Program Office at Eglin Air Force Base, Florida, was responsible for the overall project. As was common to these sorts of things, everything we did involved many meetings. So I made a lot of trips to and from the Sunshine State.

One of the key meetings that made the NDMCS possible was a gathering where every senior officer sponsoring a program gathered at a square of tables set up on a basketball court in a small gym. The top acquisition officials, the guys with the money, gave everyone a chance to defend their project in a bid to get funded. I called it "diving for dollars."

Everyone had a long story about why their project was the most deserving, the most necessary, and the earth would be knocked off its axis if their program was not funded. Some won, some didn't.

When it was Nellis's turn, I was standing behind the Range Group Commander. The top guy asked simply "what would be the result if the new microwave system was not funded?" The commander had folders full of facts, figures, statements of work, requirement details. But there was nothing that addressed such a simplistic question. He leaned an ear toward me with a "gimme an answer" look on his face. I guess the old Master Sergeant kicked in and I went for the simple and direct. "Tell him Red Flag would have to shut down," I whispered in his ear. "Red Flag will have to shut down," he repeated.

Ka-ching! "Approved", sayeth the money man. We were off and running.

Fun in the Sun

These Florida trips had their own special little adventures. Eglin Air Force Base is located in the Florida Panhandle; actually it takes up a major part of the state's western panhandle. Highway 98 along the coast passed some of the state's most attractive beaches. But our work made sure I never had a chance to take advantage of the sun and sand. That left the night life of strip clubs, bars, and sea food restaurants; but spare time was rare.

A couple of mornings on one trip provided a wonderful experience that I could not make up. My motel room overlooked an inlet instead of the beachfront. One morning I pulled the curtains back and looked across the glass-smooth water shining silver in the early morning light.

In the middle of the inlet was a guy on a sailboard, standing calmly and gliding slowly in the almost still air. A slight ripple appeared in the water next to the surfboard. No big deal, until a shiny silver-grey shape arched gently out of the ripple and back into the water. What??? This kept up as the wind surfer glided around the inlet. Then it became clear that a large dolphin was swimming along with this guy and his sailboard. Staring dumbly through the sliding glass door, this continued for quite a while. They finally glided around a bend in the shoreline and out of sight.

I rushed to get ready for work, amazed by the sight right outside my room. On the way out, I excitedly reported the morning's apparition to the motel owner. "Oh yeah, those two do that most mornings, unless the weather's bad," he said nonchalantly.

On another trip, my sister was going to be in the area on business. She was a runner and had discovered that the *Billy Bowlegs Midnight Run* would be taking place. Unable to come up with a good excuse not to take part, I agreed. A 5K fun run was part of the *Billy Bowlegs Pirate Festival*, and annual excuse for a couple days of eating, drinking, and selling cheesy souvenirs. At the stroke of midnight, a blast from a shotgun set us all off to run, in the dark, on the roads around Fort Walton Beach.

Running in what turned out to be mostly pitch black seemed a less-than-great idea. Thanks Sis. It wasn't long before my kind female sibling decided I was no challenge, said she'd see me at the finish, went into full afterburner, and rocketed off into the dark. That left me and a couple hundred new best friends to make our way through the Florida night. At one point a cry came out from up ahead, "dip in the road, dip in the road!" Getting to where the shouts were coming from I braced for an ankle-twisting ravine, only to find some moron kneeling in the middle of the road tying his shoe with runners dodging around on both sides.

I got to the finish in a not totally embarrassing time. Of course my sister was one of the top finishers and talking to the TV cameras as I puffed across the line fairly happy with my accomplishment. To recognize our participation in the event, runners were given a T-shirt with a pirate face, moon and stars, and "Billy Bowlegs Midnight Run" printed on the front. The stars and moon were phosphorescent and glowed in the dark. Nearly twenty years later the shirt was shot; but the stars and moon still glowed in the dark.

Here We Go

Rockwell Collins Transmission Systems, Richardson, Texas, had won the contract to design and install a digital microwave backbone for the range area. A trunk would go from the 554th Range Group Building to Tolicha Peak and Tonopah, with a major node installed on top of Cedar Peak. Included was a drop at Indian Springs. A Thunderbirds safety link was added as a system requirement after after a crash during practice in January 1982 killed four members of the demonstration team as they flew into the ground in a perfect diamond formation when something went terribly wrong at the bottom of a practice loop. The Thunderbirds would use the dedicated link so a safety officer at Nellis could monitor maneuvers via TV and call a break if things started going wrong.

Nevada was in the process of building a medium security prison in Indian Springs and got permission to link into one channel of our system for an emergency line to Las Vegas. During work on setting up the link I asked if it was needed in case someone escaped and they had to get a search started fast. The officer, with a straight face, said, "No. This place is so flat that if anyone gets out, we'll be able to see them for three days." Not completely accurate; but it gives a pretty good idea of what the area looked like. While there were mountains on either side, the stretch in the middle ran flat as a pancake to the horizon.

There would be a special link to TTR, which would have a modern comm/telephone system for the first time. Each mountaintop site would have individual links to communications vans on the range floor. The final stretch would be via land-line cable out to individual threats.

Get'r Done

Those were the days when working with a contractor was a pleasure. They were as dedicated to getting the job done as the Air Force was, a far cry from the way many programs go today. Rockwell Program Manager Allegra Burnworth was the key; and her project management staff was smart, creative, and dedicated. They were great to work with.

The plan called for installing a new-design, high-capacity digital microwave backbone system, a major advance over the analog links that were on their last legs. Besides a much-expanded backbone, the new system would improve the links down to the threats themselves with the goal of eventually being able to provide live video direct from the threat to Red Flag Threat Analysis to support near-real-time debriefings.

Rockwell had an extra incentive to have the program come out well. It was a chance to prove their new digital backbone design could be sized for an area as large and complex as Nellis, a near-state-sized telephone and data transmission system. It would be a win-win for both of us. It would also be a major education for me.

Digital microwave, while it sounds primitive from today's world of the Internet, fiber optic cables, satellite comms, etc.; in those days was the latest and greatest in communications technology. Learning about it was a valuable opportunity. I managed to to attend a couple of Continuing Education short courses with the Telecommunications Engineering Department at The George Washington University in Washington, DC.

Rockwell also invited me some of the Transmission Engineering Symposiums at Richardson. These were great exposure to the concepts and systems critical to the NDMCS. Even though I'd have the top digital microwave engineers at hand, once we started creating the network it was not the time for me to be guessing at how things worked. A lot was riding on what they could beat into my head, and the experts did a good job of it.

Party Time

I learned a lot, including about Texas hospitality. Every trip, the Rockwell folks went out of their way to treat me and everyone else on the trip well. Once a company executive threw a great barbecue at his pool in the Dallas suburbs. A great time was being had by all; well almost all. Being Texas, just about everyone had brought their own special chili recipe to prove who could make the best (read five-plus alarm hot!). Everyone was sitting around the pool having a grand time, sometimes setting their drinks or chili on the deck next to their seat.

Enter the owner's big, loopy Golden Lab. Over the years, Fido had developed a taste for beer and a knack for lapping it from glasses sitting next to chairs by the pool before getting shooed away, to the amusement of everyone.

He was having a grand time until coming across a bowl of hot Texas chili instead of beer next to someone's seat. The shrieks and howls were terrible, and seeing the poor animal trying to paw the fire off his tongue would have been hilarious had it not been for how sorry we felt for the poor beast. Lapping water from the pool finally made things bearably better; except for the exec who spent the next hour listening to his wife reminding him that she warned time and again about teaching the stupid dog to drink beer from glasses around the pool. She had a good point.

Following the Rules

The whole experience also demonstrated that a little common sense goes a long way in day-to-day program management. The crew from Eglin did a superior job running the NDMCS program. There were a lot of rules on the books aimed at making sure companies do not buy favor from government managers and engineers.

A good idea; but a young lieutenant from the Eglin Program Office who was handling the mounds of paperwork involved in the project personally interpreted the rules in an excessively strict straight-arrow policy for himself. At conferences and meetings, it was standard to publicly post disclaimers that the food or refreshments provided was not to be construed as an attempt to curry favor or to influence government personnel. The event planners made it clear from the start that providing lunch avoided turning the group loose to forage on their own, hoping they got back in time for the afternoon sessions. Besides, between speakers and networking the time was work, not recreation.

Unfortunately, the young man went more than a little overboard in the way he interpreted the anti-influence rules. On coffee breaks he would go down the hall to a water fountain. When the company put out sandwiches for lunch he scurried away to parts unknown, probably to do some paperwork.

While it is good to avoid untoward influence, one situation focused on little green clocks. The rental car agency we used on our trips was economical and provided good service. It had been approved for doing business with the government. There was a promotion on-going where with each rental came a small, green (the company color) ball with a digital clock face and company logo on it. They probably cost less than a buck a-piece. Digital clocks were relatively new at that time and the damn things did not work all that well.

But one day we received a scalding letter from the Program Office forbidding us from accepting any of the little clocks from the rental agency and returning the ones we had in our possession. That is just what we did. In the future, if anthropologists ever try to figure out the remains of Eglin Air Force Base, they will probably be very puzzled as to why one office would have so many little, round, green clocks that don't work.

Challenges Are Part Of The Deal

This was an interesting time, a lot of work, and some fun. But there were a few challenges worth noting.

One came as we were working on specs for the solar panels that would power the smaller communications vans set around the range floor. They would serve as nodes connecting the threats to the communications backbone. Nellis' Chief Scientist was at the solar panel company finalizing a design for how the panels would be mounted on the Porta-John-sized huts that would house the NDMCS equipment racks.

One look at the support design he faxed to me prompted a panicky phone call to catch him before the design was approved. There, right at the perfect height for scratching, two supports came together at an angle that would be too tempting for the wild horses to avoid. I could see crews having to re-align the hut and microwave shot every morning because the horses used them as scratching posts during the night, wobbling the vans and knocking the signal path out of alignment. I made it in time to implement a quick engineering change to prevent constant headaches in the future.

Let's Put This Thing In

Before the electronics could be installed, the microwave dishes, the big, bass-drum-looking antennas that would beam the signals to and from the various nodes would have to be installed; starting with the tower behind the Range Control Center.

This is where I learned that on every base there probably is some old guy who knows where all the water lines are buried, especially the ones not shown on the official blueprints. It is always wise to check with him before digging a hole anywhere. That is what I did before installing a cable conduit from the Range Control building to the tower. Some of the local "experts" were critical of the way the cable run did not go straight, but rather had what they thought was a strange dog leg behind the building. I caught a lot of grief from them over that because they said base drawings show no water lines there.

NDMCS base microwave tower   *(Photo: U.S. Air Force)*

Months later, the same crew was back to do some work, showing up with the vehicle communications engineers fear the most; a tractor-mounted back hoe. Sure enough, within an hour there was a 20 foot geyser spouting from behind the building. They nailed the water main we avoided by listening to our old civil engineer instead of the base blueprints. The boss complained loudly about how the drawings did not show a water line there. "Right," I said, pointing to the water spraying skyward and mini lake growing in the parking lot.

Dig we must   (Photo: U.S. Army Corps Engineers)

Here Comes The Stuff

We had a plan, we had a design, a truck showed up with the hardware. Now all we had to do was deliver the equipment, install it at the sites, set it up, and make sure the NDMCS worked. The contractor installation team was there, ready to go. They divided the racks of microwave radios into groups for Tolicha Peak, Cedar Peak,

Highland Peak, and the Nellis node. The bulk would be trucked to Tonopah for installation.

Everything arrived in a magnificent eighteen-wheeler with "special suspension for electronics equipment" painted on the side. A beauty of a truck it was. It would be the loveliest float in the NDMCS parade to the northland. At my prompting, it was decided that as the Chief Project Engineer I would be the one to ride in the big shiny truck.

Although I had logged many, many miles in bone-rattling Air Force trucks, up to deuce-and-a-half; never had I ridden in anything as lovely, comfortable, and spiffy. The driver was an expert on hauling electronics equipment. He pointed out the bunk compartment he and his wife used on long hauls. Bunk compartment hell, it was a small apartment. The cab and ride was far more comfortable than any car I'd ever had or would have. Listening to the driver's stories about past trips was fascinating.

It was quite a feeling to be sitting up high in this powerful luxury chariot. The first 170 miles out of Las Vegas, past the Nuclear Test Site, through Beatty and on up Highway 95 were smooth and quiet (on the inside). It was a real show watching the driver play the gearshift like a concert piano.

The bright day made cruising up the beautiful Amargosa Valley a joy. Mountains in the distance, ridges on either side, the flat floor covered with mesquite and other desert plants was, for someone who had fallen in love with the desert, like gliding through paradise. There also were the remnants of Nevada's silver mining past; old buildings, mine remnants, ghosts of a past not so long gone. Riding high in the luxurious cab was a voyage to remember.

Here Comes the Parade

At about the 170-mile mark we hit the tiny, almost-ghost-town of Goldfield. A faded memory of Nevada's silver mining days, Goldfield boasted a population of maybe a couple hundred people in those days who made their living mostly selling memorabilia and souvenirs to tourists who came up to see the old mining areas and the old jail which once was the office of Vernon Earp.

Somehow word seems to have gotten out that our convoy was coming, a few well-used blue Air Force trucks, a couple of 15-passenger vans, and us. The highway was lined with what must have been just about everybody in town, waving and cheering as we went by. For the first time all trip, the driver and I found a use for the shiny air horns on the top of the cab. It was great fun, both for us and for the Goldfield residents.

This was probably the biggest, only parade in town since the last annual Goldfield Days Parade. A marching band and some strings of beads and we could have had a Mardi Gras parade. These are the kind of things than made Nevada and its people special.

Let's Start the Install

Once we reached the range everybody was curious. Anyone who was not busy, and some who were, came over to see what we brought. The range folks were really anxious to get upgraded communications. It provided me with an extra pool of 'volunteers' to help unload what we brought.

By this time, the fiberglass huts had been set up at the various nodes around the range. The major installation would be the main site on the top of Cedar Peak. There was a cinder block building already there with the current analog equipment installed. It was a tricky but doable task to haul the NDMCS racks and electronics up there. Things would be pretty crowded until stuff could be unpacked and the circuit cards installed. But Rockwell had sent a top-notch team of installers who made the job look like child's play.

The plan called for getting the new system up and running in nine months. We could not just shut the range down for all that time, so the installation had to run in parallel with on-going operations. We would cram much of the work into a time when Red Flag was not active on the range; Maple Flag, planning conferences, time when there were no major range operations. This meant, we hoped, that the old analog system could carry the load while we installed, aligned, and checked out the new digital gear.

We installed a whole bunch of these   (Photo: Wikipedia)

207

While it was a joy to watch the crews do their thing, the heavy-duty equipment installations were accompanied by tedious checks and alignment of dozens of individual channel circuit cards. But it does not make an exciting story. In those days, a T-1 line (1.544 MB/sec) needed at least 24 circuit boards the size of a magazine, plus power supplies and control circuits. Multiply this for 28 T-1s in a backbone link and an installation consisted of one or more 19" cabinet big enough that a person could stand in. Today, this and more can be done with a box that would hold a pair of cowboy boots.

There were one or two incidents worth noting. It was about 3am at one of the comm vans on the range. Much of the work went on after range hours to avoid interfering with anything. One of the engineers was on his hands and knees adjusting channels on the bottom row of circuit boards, concentrating intensely. At that point one of the stray cows that sometimes wandered around when things were quiet stuck its head through the door and let out a loud "Moo". The poor engineer banged his head on the rack shelf and very nearly developed a concussion and laundry problem. Unfortunately for him we had no mercy and for the next several days either presented him with various cow-based trinkets or made low mooing noises when he walked by.

Up and Running, Mostly

After the system went live we found a problem. Crews would come in each morning to continue setting the ends of the nodes; but the antenna beams were out of alignment. Sometimes it was a little, sometimes they would be completely off target. No one could figure out why. There was no wind during the night, no people were around, and the occurrences were random.

Finally, one of the installation team figured it out after some clever detective work. He noticed that whenever a van's antenna would be out of align, there would be piles of horse droppings around the corners of the van. He discussed the evidence with some experienced range rats. It turned out that during the night some of the horses would come to a van and rub against the corners. Obviously, that felt so good that 'Trigger' concluded with dropping a good load to feel good all over. The range detective went around bragging that for once, when his buddies accused him of basing his conclusions on horse s**t, it was a good thing.

Some Special Days

There was one part of the operation that turned out to be a very special experience. System links had to be verified from the main mountaintop site to the individual communications vans. It is all line of sight and most were apparently clear shots; but good engineering calls for making sure and recording things officially. Verifying line of sight also provided an opportunity to collect magnetic bearings that would prove useful when it came time to align the actual transmission antennas. This was done over a weekend when absolutely nothing was scheduled on the range. That way, teams could drive around freely without having to wait for some attack or other to finish.

Although we were dealing with what was, in those days, some of the most advanced electronics on the planet, this job was accomplished with small concave mirrors with a hole in the middle. I stayed on the roof of the Cedar Peak communications site. A contractor team would drive from site to site and set up the shot. I would stand at the spots on the railing marked for the antennas that would point to particular van would go.

Coordinating over a radio, the ground crew would look through the hole in their mirror toward my site. I would watch for the sun flash from their mirror, verify that the shot was clear, and mark a compass direction. They would drive to the next site and repeat this procedure until we had all the links marked. In the afternoon, because the sun had moved to the opposite part of the sky, we repeated the procedure, this time with me looking through the hole and flashing.

What was wonderful about this was the opportunity to be alone with the animals on the range. During a workday, the wildlife usually stayed clear of humans and their noisy goings on. It would take a while for the crew to drive from one location to another. I had nothing to do but sit, look, and listen,

Normally, there was little to see and hear; but that day the peak was alive with sound. The trees were filled with birds, small animals scurried around the rocks, and larger creatures padded about the forest floor and even sniffed around the site itself. There were a couple of coyotes that spent a considerable amount of time trying to figure out how to get to my lunch they could smell through the door.

The air was cool, the sun was warm, and while sitting on the roof waiting for a radio call several little birds landed on the railing to check me out. Thanks to crumbs from my sandwiches, several of them and I became good buddies. It may sound corny; but I never knew until then what people meant when they talked about "communing with nature." I guess since then I have had a new appreciation for the world around me.

A Final Bump on the Path

As we neared the end of the installation one major site remained, Highland Peak on the northeastern corner of the range and key to the radio net for Red Flag and range missions. Things were looking good and we were ready to take on what would be the last site. Then the snow came, and came, and came. The road up could be a challenge in decent weather, with this snow it was impossible. A truck loaded with digital microwave radios and installation equipment was not going to get anywhere close to the top.

At that point, the higher-ups decided that Highland Peak would have to wait until the spring when the snow went away and trucks could make it to the top. This is where Allegra came through for the Air Force. We got our heads together, investigated local resources, and decided a helicopter was the answer. I found out that Nellis fling-wings could not be used because the system did not yet belong to the Air Force. So it was up to the contractor to come up with a solution. Besides, the base people thought it was a dumb idea. The job was running ahead of schedule and sitting around until spring would cause several months of delay and cost money, far more that renting a helicopter for a day.

I love to watch a skilled Program Manager. Allegra got my approval to try; making sure that she would not go to jail alone, I guess, should things go south. On the Nellis end, they left everything to us. She called her boss in Texas with the plan. He told her he would not approve it; whereupon she informed him that she was not asking for permission but was telling him what she was going to do and hung up. Obviously the boss knew better and did not call back. I can imagine him saying a few Hail Marys, rubbing a Star of David, and patting a Buddha statue on the belly; wondering if his career was about to come to an abrupt end.

By the next morning we had two six-foot racks sticking out the sides of a helicopter heading to the top of Highland Peak. The pilot was good; no, he was damn good. He sat the load down in the perfect spot so all we had to do was slide the crates into the main doors. It was easier than it ever would have been had we come up by truck. A couple of flights with the rest of our stuff and Highland Peak Install was ready to go.

By evening it was cold, we were tired, and we thought we lost Allegra over the side. In the middle of the night we could not find her and nearly panicked. Our thinking was that exhausted, she had accidentally gone out the back door and slipped over the edge. After some intense minutes searching every possible corner of the site we were relived to find her taking a quick nap on top of one of the equipment crates. Her gray parka had blended in with the shadows and we looked right over her in our initial search.

Finishing Up

With Highland Peak up and running, that finished the installation of the NDMCS backbone. I love it when a good plan comes together. While the backbone installation progressed, other technicians had been setting up and aligning the rest of the system. Things went very well. Every link came up with little or no trouble. Everywhere the users were amazed and pleased with the capabilities of the new system.

211

I looked forward to a major event as we officially turned the new system on. Instead, just before the planned turnover from the Contractor to the Air Force, one of the key analog links out of Nellis failed completely. Maintainers could not bring it back up no matter how hard they tried. For the first time the system did not give in to anything in their bag of tricks. The parallel digital link had just completed alignment and testing, so I OK'd switching to the new system. It worked perfectly and a possible major operational problem was avoided.

The idea caught on, so we began cutting over links as the alignments were finished. This pleased the range users very, very much; getting much more bandwidth and many more telephone channels with the long-wished-for stability.

It was strange that after months of hard work, instead of coming on line with great fanfare, everyone came in one morning and the whole Nellis Digital Microwave System was simply up and running. The last few circuits had been cut over during the night. Instead of a ceremony, the crew went to the O'Club for hamburgers and a couple of beers.

It was one of the first digital microwave systems of its kind. We brought it in significantly ahead of schedule and under budget, thanks to the hard, dedicated work of everyone involved. We formed what we called the 9-6-3 Club; scheduled for nine months, done in six months, three months ahead of schedule. In those days, everyone worked together to get the job done. The taxpayers got what they paid for, and then some.

Wrap Up

All that was left was wrapping things up. A supply of installation spares was brought in for use during the installation. Once the job was over, there was no need for them. Some had been used, others had been swapped out just to verify that what came with the racks was operating properly. Most were untouched. Rockwell could not use them. By law and ethics they could not be installed in other microwave systems or sold to other customers. It would be expensive to send them back to Texas just to be destroyed. We had to buy new spares anyhow.

Rockwell considered them a gift to the Air Force; but the rules would not allow it. The rationale was that we could use them if needed until our order of spare parts arrived. Besides, we had already paid for them in the cost of the installation contract.

This made sense to everyone except the wizards in the supply system. The rules would rather see thousands of dollars worth of perfectly good parts trashed, and the Air Force pay for an initial stock of new ones.

It was decided that some boards 'had to be destroyed', so they could not go back. Ingenious GI that I was, I agreed that they could leave the boxes with us. "We'd take care of disposing of the stuff." They thanked us for saving the company shipping costs, and we stored the boxes until somebody got around to 'trashing the stuff.'

The New Normal

From then on, the operation of the system was marked by pretty much, well, nothing. The operators were happy with extra data channels and telephone circuits that worked all the time. Data dropouts became a thing of the past and maintainers had little trouble keeping things running, thanks to the reliability of the hardware and training they'd received.

Moving On

Having just brought in a major communications backbone ahead of schedule and on budget, the chance to sit back and rest on my laurels lasted just about no time. The performance and capability of the new system not only made day-to-day range operations better, but made it possible to pursue what had been a goal since I first got to Red Flag years before – near real-time data and video from up range and same day threat debriefs.

Work on designing what would become a major link, an operational analysis and debriefing system had been on-going. The range needed a way to tie the threats into this system. I had to find a way to bring threat video down on a single voice channel. Today we watch full-motion, full-color, high-definition video on a phone that fits in our pocket. Not so in those days. At first, the systems that were available produced movement that happened in a jerky motion every quarter of a second or so. After many trips to various companies we began to find ways to transmit video that, while far from perfect, was good enough to see the pilot's basic maneuvers. Major improvements would come with time.

Another addition was a series of trips to NSA (the National Security Agency at Fort Meade, Maryland) to find a way of encrypting the backbone links to make transmission of classified data possible. These were some interesting trips to the never-never land of NSA to discuss bulk encryptors.

National Security Agency HQ   (Photo: NSA)

The engineering effort was successful. On the personal side, I received job offers for a post-Air Force career and reconnected with a dear friend from my undergraduate days in California. I turned down the job offers. That friend has been my wonderful wife for nearly three decades.

Two Old Spies

On one of my trips to NSA to discuss the bulk encryption project, I had a fascinating experience. There was a motel close to the Baltimore-Washington Parkway exit. It is where the Crypto Historical Museum and Vigilance Park sit today. Having lunch at the restaurant there one time I overheard a conversation between two old gentlemen a few tables over. One had a heavy British accent, the other German.

Vigilance Park, NSA; where the motel used to be. **(Photo: NSA)**

Although I could not make out everything they were discussing intently, it was obviously about some past activities they had been involved in. "Ah, so that's what you were doing," was a common thing I heard one or the other say after hearing out his partner.

It was not long before I figured out that here were two old spies from World War II finally hashing out some operations where they were pitted against one another, and from opposite sides! What a fascinating experience. I only wish I could have been able to better hear what they were saying, never mind that eavesdropping is impolite.

Shoot Through The Gap

During the link surveys and map studies, the team found what looked like a way to beam a microwave shot through a distant pass and probably make it possible to link the NDMCS to western sites. Such a link was not in the design, so it was not pursued. Years later, while at Ft. Irwin, California, covering a demonstration of the Army's first all-digital brigade, I discovered that a link from Nellis to Ft. Irwin and then to Edwards AFB, China Lake NAS, and other sites established the "Western Range" covering much of the Southwest had been established.

On this particular day Army units were directly talking to Air Force support pilots at Nellis and briefing mission plans. The guy I was talking to was over the moon at this ability. To say that while standing on the VIP viewing hill I felt proud would be a major understatement.

Today that's routine, in those days it was almost miraculous. Cell phones, satellites, fiber optics, and the Internet have replaced the need for most microwave systems, although they still have uses, especially if there are security issues to be addressed and complete isolation from the outside world is desirable. It is interesting to look back proudly at how what now appears so antiquated was a marvel at the time.

Down The Home Stretch

At this point, I was in what could be called the sunset of my Air Force career. I had just overseen the development of the NDMCS, the enabler for what would become the most praised part of Red Flag in the future. While my name is not attached anywhere, bringing the NDMCS on line made the Red Flag Measurement and Debriefing System (RFMDS) possible. As far as I'm concerned, it is my legacy of nearly 24 year's service. I can't help but think that everything I did, everything I experienced, somehow led up to this.

* * *

8. RFMDS

The ACMI (Air Combat Maneuvering Instrumentation) system at the Tolicha Peak Electronic Combat Range, introduced in 1975, gave birth to the RFMDS (Red Flag Measurement and Debriefing System) concept. State of the art sensors in pods carried by aircraft transmitted detailed maneuvering information to ground nodes up range in real time. This was then transmitted to Nellis where it was converted into a '3-D' picture of the dog fight and other simulated air combat for the ground controllers. This was valuable for scoring engagements and the saved data could be played back when the pilots returned to base for debriefing and training. It was a major part of Fighter Weapons School.

A pilot performs a pre-flight inspection. The ACMI pod is inboard of the missile. (USAF photo)

217

The AIS (Airborne Instrumentation Subsystem) pod was filled with sensors which gathered and transmitted flight data to the ground links. The pods were designed to be ultra-accurate (for the available technology of the day). Ground sensors used triangulation to fix an aircraft's position to within feet and information on triggering guns, firing missiles, releasing bombs was captured by the ground.

The original pods were stuffed with sensors that produced data used to create a picture of the fight. Included in the pod's instruments were Air Data Sensors, including a Pitot Tube, a Radar or Radio Altimeter, Inertial Sensor, Digital Processor and Interface, Signal Generators, Transponder, and a UHF Antenna. These sensors measured speed, altitude, attitude, G-forces, ascent/descent rate, turn and yaw rates, and roll rate. Newer versions also gathered engine power, weapons cues, weapon release point.

RFMDS Display (Photo: U.S. Air Force)

This would be communicated to the ground Control and Computation System (CCS), which in turn sent information to the Display and Debrief System (DDS) which produced a picture of the air war and the position of every aircraft in the fight. The controllers' computers and experience would be used to determine kills or misses; with 'killed' pilots given a 'time-out' in a recovery zone before being allowed back into the fight. The original ACMI generated a display that, while crude and cartoonish by today's standards, got the job done. Many arguments that once would have been settled by loud argument and hands flailing about to describe maneuvers were sorted out plain as day on the display in the ACMI at Nellis.

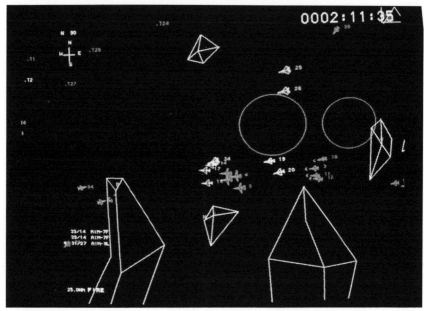

The RFMDS project was not in my portfolio. There was enough to keep me busy, like establishing standards for a new radio communications switching system for the range. At one point, I was sent to the RFMDS contractor, Cubic Defense Systems, facility where the upgrades were being developed. The meeting involved a chance to see the in-progress pod and ground station development coupled with discussions of how the just-completed NDMCS would support operations in the heat and intensity of a Red Flag exercise if the system were expanded to the northern sections of range. Just think, no more 3am wakeups.

From all indications, things would work. This proved to be true. There was ample capacity to link the ground sensors to master sites on the range and from there along the backbone to the CCS (Control and Computation System) at Nellis. This is where the engagement picture would be created for the DDS (Display and Debrief System). At this point the only unresolved issue was getting high quality video from the threats to Nellis; but a solution would come.

RFMDS Operator   (Photo: U.S. Air Force)

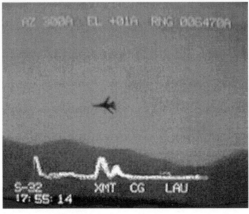

Live RFMDS threat video (Photo: U.S. Air Force)

Moving On

Not stopping, the planners took taken full advantage of technology and new techniques. Born in the analog age, the systems found that digital techniques made impossible things easy and parts smaller. The original ACMI display was like Pac-Man, with straight lines giving the shape and position of airplanes. Shots were simplistically rendered; but the result was wonderful.

Now RFMDS uses imagery from advanced CGI graphics. It could even show a simulated view from inside the cockpit. GPS was finding its way into just about everything that involved positioning information. It was a natural for this job.

RFMDS morphed into the Nellis Air Combat Training System (NACTS); more precise and comprehensive than ever. It became operational in 1999. Thanks to GPS, aircraft location was untethered from ground station triangulation and operations could be extended across the entire range. Night operations were introduced and NACTS use went beyond individual crew debriefs.

The Individual Combat Aircrew Display System (ICADS) and NACTS allowed simultaneous viewing of up to 100 aircraft and 70 electronic simulated threats at the same time, according to Cubic Defense Systems. This let commanders see "the big picture" of the war, even down into the individual cockpit while monitoring each aircraft's speed altitude, and line-of-sight in real time.

In the same release, Cubic noted that in the 25th anniversary Red Flag 01-01 (October – November 2000) pods were fitted onto 88 aircraft. Information was data-linked to Nellis, processed by the NACTS computing system and sent to 32 debriefing stations at the air range. Nine ICADS debriefing stations and six NACTS Display and Debriefing stations were in use at Nellis. (Cubic Defense Systems, *With Help from Cubic, 25th Anniversary Red Flag Provides Realistic Stealth Attack Defense Training,* Press Release via Business Wire, Nov 6, 2000)

RFMDS

RFMDS is everywhere now

In his 2012 biography of his life as an F-16CJ Wild Weasel pilot in Iraq, Lt. Col Dan Hampton shared how he became a "hero of the skies." (*Viper Pilot: A Memoir of Air Combat*) Credited with 151 combat missions, 21 hard kills on surface-to-air missile sites, 4 Distinguished Flying Crosses with Valor, and 1 Purple Heart, Col. Hampton details his path from a youngster being taught to fly by his father to graduation from the famed Fighter Weapons School at Nellis.

"The Nellis threat array is infamous for its lethality, in a training sense. In any given year, more than five hundred aircraft from all over the world fly twenty thousand training sorties against these threats. All U.S. Tactical flying units cycle through every few years, including USAF fighter squadrons based in Europe and the Far East. All the NATO air forces attend if they can afford it, and occasionally you can see Israelis and some of the friendlier Arab   (Source: *Viper Pilot* pg. 138) nations,   like Egypt and Morocco."

He described the range setup, praising the value and omni-presence of RFMDS, not just as a Red Flag asset but as an asset for Fighter Weapons School as well. "Each aircraft that flies the Nellis range carries an Air Combat Maneuvering Instrumentation (ACMI) pod. All the flight parameters and even the pilot's HUD view is fed back to a huge building containing the mission debriefing system, RFMDS, the Red Flag Mission Debriefing tactical training at Nellis. Every mission and every flight can be picked apart thanks to the ACMI pods. It's a tremendous advantage, to be able to sit with a cup of coffee at zero miles an hour and totally reconstruct a mission. All maneuvers, tactics, and every weapon that is dropped or shot is analyzed. This is how we learn, improve, evaluate, and this is another reason for American air supremacy." (Pg. 139)

Pretty much what we had in mind.

9. That's Enough

It was then that my career wound down. I'd been at Nellis a long time, and Air Force Personnel thought it was time for me to move on. My name, surprisingly, came out on the Major, Below the Zone promotion list.

The body snatchers at Randolph Air Force Base started blowing smoke up my you-know-what about how, with my background, I "was on track for command." In discussions with them they could come up with nothing more fun that what I was doing. They would not let me stay at Nellis to ride the RFMDS project through to completion and an assignment to the White House Communications Agency I'd looked at during a field trip with the Advanced Telecom course was not available. With my prior enlisted time, pinning on oak leaves would net me $20 more a month, and my "command track" looked like it would lead me straight to a communications site in the middle of nowhere on the other side of the world. I had to decide what was best. To me, the solution was simple, a trip straight to headquarters to put in my papers. The result was starting a new life full of challenge and interesting adventures, something I have never regretted. At the end of January 1985, I retired, moved across country, started a new job, and got married; all in a span of two weeks. As Mae West once said, "You only go through life once; but if you do it right, once is enough." I've never regretted my decision for a second.

Visit to an Old Haunt

It would be a little over a decade before I would get to see the results of all those efforts. While on a trip to Las Vegas to visit relatives and friends, I stopped by the Red Flag building. Once I identified myself, the Executive Officer kindly gave me a tour. The excitement, the intensity, the sense of purpose, coupled with the lingering smell of sweaty 'goat bags' was still there. The patches, the pilots, and huddles had not changed. Red Flag was still not just best in class; it was a class all its own.

When he found out when I'd been there and what I did, the colonel said "You gotta see this." The sign on the door was the same, *Threat Analysis.* When I stepped inside, though, all I could do was gasp and grab something to keep from staggering. The first thing I locked in on was a wall full of video screens with the o-so-familiar threat video, crystal clear and in full motion. Below was a series of analyst positions with a screen for video and another displaying a computerized version of the threat engagement logs I'd spent so many wee hours with "back in my day."

Then I spotted a pilot, sweaty and obviously straight from the cockpit, discussing his just-completed mission up range. All I could do was stare, and tear up a little. In my imagination, we went to the commander's office, dug the bottle of "Old Methusela" from his desk, and raised a toast to all who went before and made this possible. I went through the rest of the tour in a bit of a haze after what I'd just seen. All I could say was "Damn, I love it when a good plan comes together."

NACTS   (Photo: U.S. Air Force)

10. Red Flag, Present and Future

Red Flag and its spinoffs, Maple Flag, and Red Flag – Alaska, continue to be the premier air combat training on the planet. While there is occasional criticism that the exercises reflect a combat environment of the past, it is always easy to bad-mouth the best. Planners continue to build, operators continue to fine tune, the exercises continue to be the best in the world.

The following are the latest fact sheets released in February 2012 by the Air Force on Red Flag, the Range Complex, and Red Flag – Alaska. Compared to my days there, things have come a long way. Red Flag continues to march forward; still based on a solid foundation of excellence, dedication, and creativity. When it comes to fighter pilots, it produces the best by being the best.

New technology makes continued change and upgrades possible. The range has been re-worked countless times to make practice for all manner of missions possible. One of the first was re-building the target and threat array to replicate what pilots would face during the First Gulf War. These changes continued as the force found new challenges facing it.

The following are the latest official Air Force releases on these exercises today.

414th COMBAT TRAINING SQUADRON "RED FLAG"

U.S. Air Force Fact Sheet

RED FLAG, a realistic combat training exercise involving the air forces of the United States and its allies, is coordinated at Nellis Air Force Base, Nev., and conducted on the vast bombing and gunnery ranges of the Nevada Test and Training Range. It is one of a series of advanced training programs administered by the United States Air Force Warfare Center and Nellis and executed through the 414th Combat Training Squadron.

RED FLAG was established in 1975 as one of the initiatives directed by General Robert J. Dixon, then commander of Tactical Air Command, to better prepare our forces for combat. Tasked to plan and control this training, the 414th Combat Training Squadron's mission is to maximize the combat readiness, capability and survivability of participating units by providing realistic training in a combined air, ground, space and electronic threat environment while providing for a free exchange of ideas between forces.

Aircraft and personnel deploy to Nellis for RED FLAG under the Air Expeditionary Force concept and make up the exercise's "Blue" forces. By working together, these Blue forces are able to utilize the diverse capabilities of their aircraft to execute specific missions, such as air interdiction, combat search and rescue, close air support, dynamic targeting and defensive counter air. These forces use various tactics to attack NTTR targets such as mock airfields, vehicle convoys, tanks, parked aircraft, bunkered defensive positions and missile sites. These targets are defended by a variety of simulated "Red" force ground and air threats to give participant aircrews the most realistic combat training possible.

The Red force threats are aligned under the 57th Adversary Tactics Group, which controls seven squadrons of USAF Aggressors, including fighter, space, information operations and air defense units. The Aggressors are specially trained to replicate the tactics and techniques of potential adversaries and provide a scalable threat presentation to Blue forces which aids in achieving the desired learning outcomes for each mission.

A typical RED FLAG exercise involves a variety of attack, fighter and bomber aircraft (F-15E, F-16, F/A-18, A-10, B-1, B-2, etc.), reconnaissance aircraft (Predator, Global Hawk, RC-135, U-2), electronic warfare aircraft (EC-130s, EA-6Bs and F-16CJs), air superiority aircraft (F-22, F-15C, etc), airlift support (C-130, C-17), search and rescue aircraft (HH-60, HC-130, CH-47), aerial refueling aircraft (KC-130, KC-135, KC-10, etc), Command and Control aircraft (E-3, E-8C, E-2C, etc) as well as ground based Command and Control, Space, and Cyber Forces.

A "White" force in RED FLAG uses the Nellis Air Combat Training System (NACTS) monitor this mock combat between Red and Blue. NACTS is the world's most sophisticated tracking system for combat training exercises and allows commanders, safety observers and exercise directors to monitor the mission and keep score of simulated 'kills' while viewing the simulated air battle as it occurs.

As RED FLAG expanded to include all spectrums of warfare (command, control, intelligence, electronic warfare) and added night missions to each exercise period, the combination of NACTS, improved tactics, and increased aircraft/aircrew capabilities improved flying safety.

All four U.S. military services, their Guard/Reserve components and the air forces of other countries participate in each RED FLAG exercise. Since 1975, 28 countries have joined the U.S. in these exercises. Several other countries have participated as observers. RED FLAG has provided training for more than

440,000 military personnel, including more than 145,000 aircrew members flying more than 385,000 sorties and logging more than 660,000 hours of flying time.

This mock battle in the skies over the Nevada Test and Training Range has yielded results that will increase the combat capability of our armed forces for any future combat situation.

(Current as of February 2012)

99th Air Base Wing Public Affairs

Nellis Air Force Base, Nevada

Nellis Range Complex, Nellis Air Force Range (NAFR)

The 99th Range Group (ACC) operates, maintains, and develops four geographically separated electronic scoring sites, an instrumentation support facility, and the 3.1-million-acre Nellis Range Complex, including two emergency/divert airfields. It formulates concepts and advocates requirements to support Departments of Defense and Energy advanced composite training, tactics development, electronic combat, testing, and research and development. Nevada Test and Training Range (NTTR) is formally referred to as the Nellis Air Force Range.

The 99th Range Squadron (ACC) operates, maintains, and develops the Nellis Range Complex, comprising 3.1 million acres and 12,000 square miles of airspace. It supports advanced composite force training, tactics development, and testing. It coordinates operational and support matters with MAJCOMs, Departments of Defense, Energy, and Interior, as well as other federal, state, and local government agencies to meet a broad spectrum of range user requirements. The 99th Range Squadron commands two detachments: Indian Springs Air Force Auxiliary Field, which manages Nellis' Southern Ranges, and Tonopah Test Range Airfield, which manages Nellis' Northern Ranges.

The 99th Range Support Squadron (ACC) operates, maintains, and develops four geographically separated electronic scoring sites at Belle Fourche, SD; La Junta, CO; Dugway, UT; and Harrison, AR, as well as an instrumentation support facility located at Ellsworth AFB, SD. It coordinates operational and support matters to meet advanced composite force training, tactics development, electronic combat, and testing program requirements. It acquires and manages contract support for operations, maintenance, instrumentation, and communications systems.

The Nellis Range Complex (NRC) has been designated a Major Range and Test Facility by the Department of Defense (DoD), providing threat simulators in combat-like environments. The 99th Range Group also acts as the ACC lead range advocate to provide centralized expertise for the development of ACC test and training ranges as directed by HQ ACC. The Range Group carries out its charge through the efforts of some 600 contractors and nearly 300 military and civil service personnel. In short, the 99th Range Group is a team of professionals providing the world's premier integrated battlespace environment.

The five geographical areas of the Nellis Range Complex consist of: Restricted Areas R-4806, primarily used for testing and munitions training; R-4807, used for electronic combat and munitions training; R-4808, used by the Nevada Test Site; R-4809, used primarily as an electronic combat range; and the Desert Military Operating Area, used for air-to-air training.

The Nellis Range Complex is located between Las Vegas and Tonopah in Southwestern Nevada and consists of five adjacent geographical areas. The ground is mostly barren, consisting mainly of flat, dry lake beds, dry washes, desert vegetation, and rugged, mountainous terrain. The land occupied by the NRC is more than 3.1 million acres, combined with more than 12,000 square miles of airspace. The 99th Range Squadron, which controls the range, is located on Nellis Air Force Base, approximately eight miles northeast of Las Vegas.

The NRC overlays large portions of Clark, Lincoln, and Nye counties in southern Nevada and small portions of Iron and Washington counties in southwest Utah. Land uses in this area include the military land use area of the Nellis Air Force Range [NAFR].

The NAFR consists of approximately 3 million acres. The majority of the NAFR consists of lands withdrawn from the BLM. Withdrawn lands refers to land which is set aside for a specific use. In this case, it is land that has been set aside for military use that is not available for public use. It remains under the jurisdiction and management of the agency that is responsible for the land. The Air Force must comply with all uses, policies, programs, federal requirements as mandated and administered through BLM. The 389,420-acre Nevada Wild Horse Range is included in the NAFR and is administered by the BLM. Approximately 816,400 acres of the NAFR have been withdrawn from the Desert National Wildlife Range (DNWR). The Air Force and USFWS jointly manage this area. The Nevada Test Site, administered by the DOE, is contiguous with the NAFR in the southwestern part of the NRC. Public access to the NAFR and the Nevada Test Site is highly restricted, although some areas support grazing leases. The NAFR is used for training, testing, and weapons evaluation operations for the Air Force, Army, Marine Corps, National Guard, Navy, DOE, and reserve forces. Target complexes with bombing circles and triangles, and simulated runways, airfields, and convoys are situated on parts of the NAFR.

The major land uses beneath the remainder of the NRC area are managed by the BLM and are primarily used for the production of cattle and other livestock. This rural area is scattered with widely separated small communities, farms, and ranches. Limited private land area also occurs within this portion of the NRC. Communities within the area include Pioche (population approximately 800), Alamo (400), and Caliente (1,100) within Nevada, and Modena (35) in Utah (Rand McNally and Company, 1996; U.S. Air Force, 1994a). Portions of the Humboldt National Forest in Nevada and the Dixie National Forest in Utah are also situated within this area. Some areas are controlled by the state of Nevada, including several state parks (Beaver Dam, Cathedral Gorge, and Echo Canyon).

State parks and BLM recreational sites support recreational land uses. The Humboldt National Forest area within the boundaries of the NRC includes the Quinn Canyon and Grant Range wilderness areas. Approximately 18 wildlife resource areas and national wildlife refuge (NWR) system units are either totally or partially beneath the NRC. These areas are administered by three agencies: the USFWS manages approximately 1.26 million acres, the U.S. Forest Service manages approximately 57,000 acres, and the BLM manages 927,503 acres, totaling approximately 2.24 million acres, or approximately 17 percent of the total NRC. The two major NWR system units are the DNWR, partially overlapping with the NAFR and the Pahranagat NWR.

The Nellis Range Complex maintains the most realistic integrated threat simulator environment in the world. In addition to the wide assortment of SAMS, AAAs, and acquisition radars operated by Range Squadron personnel from 39th Intelligence Squadron, maintain and operate a variety of radar and communications jamming equipment. Coupled with the Nellis Red Flag Measurement and Debriefing System (RFMDS), these assets provide superior year-round training to U.S. and allied aircrews in both competition and training exercises.

Should real-world contingency require certain configurations for training, targets can be built or changed quickly. During the Persian Gulf war, for example, range contractors transformed a runway configuration from a typical former Warsaw Pact country's configuration to one based on what allied aircrews would see in Iraq using data gathered from intelligence reports and photo reconnaissance missions.

The Nellis Range Complex supports numerous Red Flag and Green Flag Exercises and USAF Weapons School exercises each year. In addition, a number of threat simulators are deployed to Cold Lake, Canada, once a year in support of Maple Flag. The NRC also hosts the Gunsmoke competition every two years. Operational Testing and Evaluation missions on the NRC are supported by upgraded Television Ordnance Scoring Systems

(TOSS) and state-of-the-art Kineto Tracking Mount documentation and Time Space Position Information (TSPI) data. Additional capabilities include support for: Operational Flight Programs (OFP), Qualification Operational Test & Evaluation (QOT&E), Tactics Development & Evaluation (TD&E), and Flow-on Test & Evaluation (FOT&E).

Detachment 1 of the 99th Range Squadron is responsible for supporting all ACC activities at Indian Springs Air Force Auxiliary Field and the Southern Ranges of the NAFR. They direct support of DoD, Department of Energy (DOE) research, development, and testing programs. The detachment also supports recovery of emergency/divert military aircraft involved in major aircrew training exercises, such as Red Flag.

The ranges offer a wide variety of targets for inert and live munitions for test and training missions. Examples of missions performed on the Southern Ranges include strafing and employment of cluster bomb unit drops, aircraft-mounted rockets, laser-guided bombs, and general-purpose bombs. Although various forms of testing are done throughout the NAFR, Det-1's Range 63 is configured to provide real-time data for operational testing and evaluation missions. This is accomplished through a variety of means, including upgraded Television Ordnance Scoring Systems (TOSS), state-of-the-art Kineto Tracking Mount optical documentation, ballistics data reduction, and Time Space Position information (TSPI) data.

Detachment 2 of the 99th Range Squadron is responsible for, and directs, all ACC activities at Tonopah Test Range Airfield and the Northern Ranges. Like their southern partners, the detachment directs support of DoD, DOE research, development, and testing programs and also supports recovery of emergency/divert military aircraft involved in major testing and aircrew training exercises. The Northern Ranges offer unique test and training targets such as airfields, missile sites, trains, and bunker formations and a wide variety of threat simulators, uniquely tailored to individual mission requirements. Det-2's mission includes providing sophisticated training, testing, and weapons evaluation for various defense and other federal agencies, as well as allied nations.

To support aircrew training and testing, the Northern Ranges are further divided into the Tonopah Electronic Combat Range and the Tolicha Peak Electronic Combat Range. The detachment coordinates operational and support matters with Department of Interior, Bureau of Land Management, and other federal, state, and local government agencies. Within its boundaries, the Northern Ranges include the Nevada Wildhorse Range--the first wild horse area established in the United States. A superb host-tenant relationship exists between Det-2 and Sandia National Laboratories, which operates a specific portion of the Tonopah Test Range.

The NAFR is one of, if not the most, sophisticated, versatile, and complex training and test range in the United States. The often varied and complex nature of the NAFR represents many safety challenges that are addressed on a daily basis at Nellis AFB. Safety considerations are addressed in the early planning stages of test and training missions as well as in the daily operations of the personnel who constantly access the range, either in the air or on the ground. During the planning stages of test and training missions, the users of the NAFR are required to coordinate their mission with the 99th Range Squadron to determine the need for a Range Safety Approval (RSA) and availability of range facilities and airspace. Tests and selected training exercises involving armament, weapons delivery systems, or laser systems missions not previously approved required an RSA signed by the commander. The RSA is prepared by the Air Warfare Center's Range Safety Office (HQ AWFC/SEY). The preparation of the RSA involves a team effort between the user, the range squadron, and the Range Safety Office and results in the identification of all significant hazards associated with the mission as well as the assignment of an overall risk rating. Any RSA assigned a risk rating greater than an acceptable or low rating will identify the reason for the moderate or high risk rating as well as a recommendation to the range squadron commander for approval or disapproval. The range squadron commander is responsible for assessing the risks and accepting or rejecting those risks associated with test or training operations.

The original Nellis Air Force Range was established by President Roosevelt in 1940 and was originally referred to as the Las Vegas Bombing and Gunnery Range. It consisted of nearly 3,560,000 acres. In 1942, Executive Order 9019 returned approximately 937,730 acres to the control of the Department of the Interior. Approximately 554,037 acres of this land was part of what was to make up Area A. This portion of Area A, which came from the former Las Vegas Bombing and Gunnery Range, had no specified designated targets. Later, in 1953, the Tonopah Bombing and Gunnery Range relinquished approximately 154,584 acres to the Department of the Interior. These two tracts of land comprise Area A, approximately 708,621 acres. There was no documentation found to identify specific target locations for the land which came from the Tonopah Bombing and Gunnery Range.

The former Nellis Air Force Range, Area A, is located in Lincoln and Nye Counties north and northeast of the present-day boundaries of the Nellis Air Force Range complex north of Las Vegas, Nevada. The site consists of approximately 708,621 acres. Although the majority of the area is used for wildlife conservation and is controlled by the Bureau of Land Management (BLM), some areas are leased to ranchers for cattle grazing. The remainder of the land is owned by various private landowners using the land for cattle ranching and farming.

The majority of Area A is still used today as a Military Operations Area (MOA) for flyovers by the pilots from Nellis Air

Force Base. This area consists of airspace use only and is not part of the active firing ranges of Nellis Air Force Range.

RED FLAG – ALASKA
U.S. Air Force Fact Sheet

Eielson Air Force Base and Joint Base Elmendorf-Richardson are the home of RED FLAG-Alaska. Each exercise is a joint/coalition, tactical air combat employment exercise which corresponds to the operational capability of participating units. In other words, exercises often involve several units whose military mission may differ significantly from those of other participating units. RED FLAG-Alaska planners take these factors into consideration when designing exercises so participants get the maximum training possible without being unfairly disadvantaged during simulated combat scenarios.

RED FLAG-Alaska is a Pacific Air Forces-sponsored, Joint National Training Capability accredited exercise Originally named COPE THUNDER, it was moved to Eielson Air Force Base, Alaska, from Clark Air Base, Philippines, in 1992 after the eruption of Mount Pinatubo on June 15, 1991 forced the curtailment of operations. COPE THUNDER was re-designated RED FLAG-Alaska in 2006.

Red Flag: Past, Present & Future

When the decision was made to relocate COPE THUNDER, Air Force officials viewed Eielson as the most logical choice. That decision was based partly on the fact that Eielson's 353rd Combat Training Squadron already controlled and maintained three major military flight training ranges in Alaska.

Initiated in 1976, COPE THUNDER was devised as a way to give aircrews their first taste of warfare and quickly grew into PACAF's "premier simulated combat airpower employment exercise."

Prior to Operation Desert Storm, less than one-fifth of the U.S. Air Force's primary fighter pilots had seen actual combat. While the percentage of combat-experienced pilots has increased in recent years, at the time, a high percentage of pilots had no combat experience. Analysis indicates most combat losses occur during an aircrew's first eight to 10 missions. Therefore, the goal of RED FLAG-Alaska is to provide each aircrew with these first vital missions, increasing their chances of survival in combat environments.

RED FLAG-Alaska participants are organized into "Red" aggressor forces and "Blue" coalition forces. "White" forces represent the neutral controlling agency. The Red force includes air-to-air fighters, ground-control intercept, and surface air defense forces to simulate threats posed by potentially hostile nations. These forces generally employ defensive counter-air tactics directed by ground-control intercept sites. Range threat emitters -- electronic devices which send out signals simulating anti-aircraft artillery and surface-to-air missile launches -- provide valuable surface-to-air training and are operated by civilian contractors as directed by 353rd Combat Training Squadron technicians. The Blue force includes the full spectrum of U.S. and allied tactical and support units. Because the Red and Blue forces meet in a simulated hostile, non-cooperative training environment, the job of controlling the mock war and ensuring safety falls to the White neutral force.

On average, more than 1,000 people and up to 60 aircraft deploy to Eielson, and an additional 500 people and 40 aircraft deploy to Joint Base Elmendorf-Richardson, for each RED FLAG-Alaska exercise. Most participating RED FLAG-Alaska units arrive a week prior to the actual exercise. During that time, aircrews may fly one or two range orientation flights, make physical and mental preparations, hone up on local flying restrictions, receive local safety and survival briefings, and work on developing orientation plans.

During the two-week employment phase of the exercise, aircrews are subjected to every conceivable combat threat. Scenarios are shaped to meet each exercise's specific training objectives. All units are involved in the development of exercise training objectives. At the height of the exercise, up to 70 jet fighters can be operating in the same airspace at one time. Typically, RED FLAG-Alaska conducts two combat training missions each day.

All RED FLAG-Alaska exercises take place in the Joint Pacific Range Complex over Alaska as well as a portion of Western Canadian airspace. The entire airspace is made up of extensive Military Operations Areas, Special Use Airspace, and ranges, for a total airspace of more than 67,000 square miles.

Since its inception, thousands of people from all four branches of the US military, as well as the armed services of multiple countries from around the world, have taken part in RED FLAG-Alaska and Cooperative COPE THUNDER exercises. Last year, more than 5,000 people deployed to RED FLAG-Alaska, and participating aircrews flew over 4,000 missions.

Aircrews aren't the only ones who benefit from the RED FLAG-Alaska experience. Exercises provide an operations training environment for participants such as unit-level intelligence experts, maintenance crews, and command and control elements.

By providing generic scenarios using common worldwide threats and simulated combat conditions, RED FLAG-Alaska gives everyone an opportunity to make the tough calls often required in combat.

RF-A executes the world's premier tactical joint and coalition air combat employment exercise, designed to replicate the stresses that warfighters must face during their first eight to ten combat sorties. RF-A has the assets, range, and support structure to train to joint and combined war fighting doctrine against realistic and robust enemy integrated threat systems, under safe and controlled conditions.

Washington Causes Red Flag Cancellation

The exercise planned for July 2013, which was canceled because of spending cuts mandated by Washington's boneheaded budget sequester, would have included participants from Australia and Great Britain. This resulted in a loss of roughly 2,500 training flight hours from the roughly 125 aircraft scheduled.

Bibliography

3380th Technical School, United States Air Force, Biloxi Photo Company, Biloxi, Mississippi, 1961.

Boyne, Walter J., *Red Flag: The World Famous Training Exercise is 25 Years Old This Month,* Air Force Magazine, November 2000.

CIA, *The Central Intelligence Agency and Overhead Reconnaissance: The U-2 and OXCART Programs, 1954 – 1974, 1992,* approved for release 2013/06/25, through the George Washington University Security Archives program.

Hall, George, *Superbase Nellis, The Home of Red Flag,* Osprey Publishing Ltd., London, U.K., 1988.

Hampton, Dan, *Viper Pilot: A Memoir of Air Combat,* William Morrow, an imprint of HarperCollins Publishers, 2012.

Jacobsen, Annie, *Area 51: An Uncensored History of America's Top Secret Military Base,* Little, Brown and Co., New York, New York, 2011.

Nalty, Bernard C, *Tactics and Techniques of Electronic Warfare,* Defense Lion Publications, Newtown, Connecticut, 2013

National Fire Protection Association Report, *Investigative Report on the MGM Grand Hotel Fire,* Las Vegas, Nevada, Nov 21, 1980.

Reyno, Mike, *Maple Flag,* Concord Publishing Co., New Territories, Hong Kong, 1992.

Bibliography

Skinner, Michael (Photography by George Hall), *RED FLAG: Air Combat for the '80s,* the Presidio *Airpower* Series, Presidio Press, Novato, California, 1984.

Skinner, Michael (Photography by George Hall), *RED FLAG: Air Combat for the '90s*, Motorbooks International Publishers and Wholesalers, Osceola, Wisconsin, 1993.

Wiene, Roger, *Red Rover: Inside Story of Robotic Space Exploration, from Genesis to the Mars Rover Curiosity*, Basic Books, Perseus Books Group, New York, New York, 2013.

Winkler, David F., *Searching the Skies: The Legacy of the United States Cold War Defense Radar Program*, USAF Air Combat Command, 1997.